D1436756

D1436757

250 564 186

The ILLUSTRATED HISTORY _of_
FIRE ENGINES

The ILLUSTRATED HISTORY _of_
FIRE ENGINES

KEITH RYAN

Eagle
Editions

A QUANTUM BOOK

Published by Eagle Editions Ltd.
11 Heathfield
Royston
Hertfordshire SG8 5BW

Copyright ©MCMXCVIII
Quintet Publishing Ltd.

This edition printed 2003

ISBN 1-86160-519-6

QUMFEN

This book is produced by
Quantum Publishing Ltd
6 Blundell Street
London N7 9BH

Printed in Singapore by
Star Standard Industries (Pte) Ltd.

Contents

1 Where There's Smoke

LIGHTNING STRIKES A TREE IN A STORM, BRUSH FIRES BURST IN SUMMER HEAT, ROCKS SCRAPE TOGETHER AND SET OFF A SPARK – FIRE IS BORN AND WE ARE MESMERIZED. FIRE HAS FILLED OUR MEMORIES FROM OUR VERY FIRST ENCOUNTER, REPRESENTING A POWER WE HAVE TRIED TO MASTER. IN OUR MYTHOLOGY FIRE IS AN UNWILLING GIFT OF THE GODS, SOMETHING TO BE FEARED, RESPECTED, AND CONTROLLED. OVER CENTURIES WE HAVE BUILT MACHINES TO ACHIEVE THIS CONTROL BUT, FROM THE DRY SUMMER FIRES BLAZING ACROSS THE CALIFORNIAN HILLS TO THE WILD AFRICAN BRUSH FIRES, IT REMAINS A CONSTANT, UNPREDICTABLE POWER.

The Dawn of Fire

The history of fire engines begins with our earliest recorded history, a web of invention and innovation describing our efforts to stop fire in its tracks. Today, we can anticipate a fire's arrival and predict its movement. It powers our engines, clears the grasslands to allow for new growth and warms our homes. One of the greatest threats to our safety has become one of our greatest assets.

Mythology paints a clear picture of our relationship with fire, from the burning Phoenix rising from its own ashes to fire-breathing dragons and gods of Japanese and Celtic myths. At the heart of each fiery story is locked our own fear of the flame's destructive power intertwined with respect for its potential.

The Illustrated History of Fire Engines examines the inventions and strategies marking the evolution of our relationship with fire. From simple hand pumps and bucket brigades to contemporary fire fighting technology, this book considers the significant stages of fire engine development, the machines, their builders, and the battles they have fought.

AT THE HEART OF THE FLAME

A fire has simple needs. It requires fuel, oxygen, and heat – the so-called "Fire Triangle" – to survive. If one of these three is contained a fire dies out: water poured over a fire absorbs the heat of the fuel being burned and extinguishes the fire; steam produced by the boiled water uses up extra oxygen and smothers the fire; if there is no remaining fuel to burn, a fire will effectively burn itself out.

While the basic elements of fire have not always been understood, most of the earliest fire fighting techniques have related directly to the Fire Triangle. Water was observed "putting out" fires, so most early fire fighting was concerned with dousing flames. Even today, engines are designed to move the greatest quantities of water possible. Smothering a fire was an alternative, though the first wielders of water-soaked fire fighting mats may not have realized exactly how they were killing the flames. Tearing down the buildings surrounding a fire in order to stop it from spreading was another option, removing the fuel from the fire and allowing it to "burn out."

Current understanding of the life of a fire, however, has led to innovative solutions such as foam and chemical fire fighting apparatus that smothers the flames, helicopter-borne infrared sensors that detect where a fire may spread based on concentrations of heat, and great tarps that create a vacuum around a fire scene. Engines today are massive machines with enough equipment to put out a fire, rescue trapped fire victims, and treat the injured. Fire fighters are experts in their field and fighting a fire has become an organized event. These developments may be impressive and scientifically sound but there is something very romantic about early fire fighters rushing headlong into a blaze, water buckets and hooks in hand, ready to do battle with one of our most basic enemies.

FIGHTING FIRE THE OLD-FASHIONED WAY

Practical fire engines were not built until the mid-seventeenth century and there is little recorded information on these machines. The history of fire engines goes back even further. The ideas that led to the first fire engines go as far back as ancient Rome, passed on from country to country over centuries. It began with sentries watching for smoke, a method carried from Rome into medieval England which eventually crossed the Atlantic into colonial America. These societies held a common fear of fire – experience proved that their best weapon was vigilance. The most effective preventative measure was a network of scouts. When a fire broke out, the standard response was to pull down the surrounding buildings to stop it spreading, not putting out the flames. It was a sloppy and often unreliable way to work but it was the best that could be done. It also encouraged inventiveness and creativity in fire fighting strategies, setting the stage for later fire engine innovation.

Many fire fighting techniques in Europe and, later, America were based on the strategies developed in Rome during the third and fourth centuries BC. Roman society's concern about fires probably relates to the wealth accumulated by the Empire. In response, the "public family" (*Familia Publica*) was created, an organization of slaves who watched for and responded to any outbreaks of fire. There wasn't enough confidence in slaves to keep them in the fire fighting business for long – no-one believed they would keep the public's best interests at heart – which prompted the eventual creation of the "vigilants" (*Vigilis*). Emperor Augustus established this organization, which eventually numbered in the thousands and protected the city for over 500 years. It was first made up of freed former slaves but, as its status in the community grew, citizens began to join. It is the earliest known example of a paid fire fighting organization, generating the first debates over paid fire fighting professionals versus volunteer fire fighters.

The fire fighting apparatus was rudimentary at best but represented equipment used for centuries to come in Rome, Europe, and the Americas. It is described in the works of some of the best-known writers of the time, including Vitruvius, Pliny, and Hero. Equipment included hooks and poles to pull down walls, woven mats soaked in water, buckets, and simple but effective water pumps.

In his book *On Architecture*, the Roman architect and engineer Vitruvius speaks about one of the most successful early pump engine builders, an Alexandrian named Ctesibius who created a simple but vital engine that held the key to successful and more powerful fire engines to come: an air chamber.

The Ctesibius pump had all the elements of a manual fire engine pump long before they came into popular use. Vitruvius described it as a small brass container enclosing two cylinders with valves and greased pistons, topped by levers for pumping the pistons and a jointed pipe through which water was forced via the air chamber. The chamber created substantial pressure and projected a strong jet of water. As the levers were pumped, water was drawn into the chamber and the air compressed to create greater atmospheric pressure within the chamber than outside. While the

pumps continued pumping, the pressure inside the chamber forced water out and drew in more water as the pressure equalized. With each stroke, water was being brought in and sent out in a consistent and regular rhythm. This simple pump would prove one of the most important developments in fire engine history. The basic components used by Ctesibius would reappear over 1,600 years later in Europe as part of the earliest successful manual fire engine pumps.

Ctesibius was not the only pump builder. The Roman historian Hero described a double-piston pump in the second century BC, very similar to Ctesibius', while the *Fifty Letters of Pliny* mentions a complete fire brigade operation as well as a "sipo, nulla hama," thought to be yet another manual water pumping system. These brigades and their pumps were very effective at fighting fires – so

effective that their methods carried over into England and the rest of Europe. Unfortunately, the air chamber pumping system did not make the original journey

Pump engines were used in ancient Rome but most of the fire fighters' tools remained basic for centuries: poles and hooks to create fire breaks between buildings, ladders to reach rooftops, and ever-popular leather buckets to carry water. These tools were a desperate attempt to stop a fire spreading, not to defeat the fire. At their best, early fire fighting efforts were based on strength and willpower, not technology. Once a blaze was underway, everyone was expected to pitch in, a philosophy maintained and sometimes even required by law. As the name of the ancient fire fighting brigade, *Familia Publica*, suggests, fighting fires involved everyone in the "public family."

PATERSON FIRE ASSOCIATION.

NEW-JERSEY.

THESE ARE TO CERTIFY:

That *Nathaniel Lowe*

is, pursuant to Law, nominated and appointed one of

the **FIREMEN** of the Town of **PATERSON**.

By the Wardens.

 Secretary.

Paterson, *July 1st 1838*

Day & Burnett, Printers, Paterson.

OPPOSITE PAGE:

A Paterson (New Jersey) Fire Association certificate, dated July 1, 1828, naming Nathaniel Lane an official fire fighter.

LEFT: **Hand rattles used in early colonial America were brought from England, used to wake up sleeping citizens. These were the source of the infamous 'Rattle Watch' name given to the scouts.**

AN OUNCE OF PREVENTION

As the Roman Empire spread, it brought its fire fighting strategies to England and Europe, where they were used for well over a thousand years. Instead of advancing fire engine technology that was developed under the Empire, though, the focus was on fire prevention. William the Conqueror introduced fire safety measures in England in the early years of the eleventh century AD, ordering the public to put out their fires at night ("couvre feu" or "fire cover" – the modern "curfew"). This was not a popular move – starting a fire was a time consuming process and it was easier to keep a small flame constantly alight. Legislation was also introduced as early as the twelfth century to relieve some of London's overcrowding and limit the wooden buildings being built throughout the very dense city.

Roman influence lasted well into the sixteenth century. Fire prevention was all-important while the development of fire engines remained stagnant. In England in particular, the same apparatus was being used (hook, pole, ladder, and bucket) as had been used by the Vigilis almost two thousand years before, but the air chamber pump had slipped out of use.

Prevention meant strict rules: building codes encouraged non-flammable construction materials (chimneys had been made of wood for years) and legislation forced citizens to buy fire-fighting equipment – in some cases, more than one bucket was ordered, so great was the fear of shortage. If the alarm sounded, people were expected to come running, bucket in hand, to join the bucket brigade and take on the fire. Vigilance was considered the best response to the threat of fire in England, just as it had been in ancient Rome. Fire engines, especially Ctesibius's air chamber, continued to gather dust.

As the ancient Romans had influenced Europe, so Europeans brought their ways to the New World, demonstrated in the records of Dutch-controlled New Amsterdam (known to us today as New York)

as late as the seventeenth century. As in ancient Rome and medieval Europe before, fire engines were not as important as fire prevention. Anything that made the citizens' lives safer or at least protected their property was welcomed, prompting Governor Stuyvesant to institute the first fire prevention legislation to the New World. Wooden chimneys and thatched roofs were abolished, chimneys were swept regularly, fire marshals were assigned, and, perhaps the most famous of all, the Rattle Watch was born. Augustine Costello explains in his work, *Our Fireman: A History of the New York Fire Departments (Volunteer and Paid)*, first published in 1887:

> The burning of a small log house on a bluff overlooking the bay … led to the establishment of the first fire company in 1658. This organization, disrespectfully dubbed the "Prowlers," consisted of eight men, furnished with two hundred and fifty buckets, hooks, and small ladders, and each of its members was expected to walk the street from nine o'clock at night until morning drum-beat, watching for fire while the town slumbered. This company was organized by ambitious young men, and was known also as the "rattle watch." It was soon increased to fifty members, and did duty from nine P.M. until sunrise, all the citizens who could be roused from their beds assisting in case of fire.

If a fire broke out, the men of the Watch would spin loud rattles to wake up the nearby residents, giving the group its distinctive name. Each citizen had to "fill three buckets with water after sunset, and place them on his doorstep for the use of the fire patrol in case of need," and 10 buckets were filled at the town pump and left on a rack for the Watch, just in case. Though they seem obvious today, these measures offered peace of mind to a community that had been plagued by fire. They also revived the idea of a structured body organized specifically to fight fires – the Vigilis of ancient Rome, in addition to fire fighting duties, was responsible for the recapturing of runaway slaves as well as watching over the clothes of bathers at the public baths.

ABOVE: **Leather buckets were used for centuries in the front line of fire fighting efforts in the United States and around the world.**

OPPOSITE PAGE: **Mobile and functional, this hand pump was designed for use against smaller, contained fires.**

When New Netherland was taken from the Dutch by the British in 1664, New Amsterdam was among the territory claimed and was renamed New York after the Duke of York, brother to Charles II. The new government was as concerned about fire prevention as its predecessor and instituted measures to improve fire fighting techniques. Anyone holding on to one of the city's buckets, hooks, or ladders had to return them to the mayor immediately; wells were dug in key areas of the city; and proper chimneys were built where none were present. A city chimney sweep patrolled the houses and cleaned the blocked flues. On hearing the cry, "Throw out your buckets!" and the bells of the city's fort and church steeples, citizens threw their three-gallon leather buckets into the street. These were collected by the men rushing to the fire – usually the homeowner from each household, according to custom. Two solid lines of citizens stretched from the water source to the fire, one to bring the water, the other to return the empty buckets for more. According to Costello, "No one was permitted to break through those lines, and if they attempted to do so, and would not lend a helping hand, a bucket of water or several were instantly thrown over him."

Early fire fighting techniques were limited by the resources at hand: the quantity and quality of equipment, the number of people available to help battle a blaze, and – the most vital resource of all – luck. Some fires were fought solely by prayer in medieval Europe – unfortunately, most burning buildings didn't have a prayer and were written off from the first sign of smoke. The arrival of a functional manual hand pump revealed a light at the end of the tunnel. Even with only one or two rudimentary machines, citizens often were willing to work that much harder to stop a fire. For almost two thousand years, people had been fighting fires through pure willpower. Fire fighting strategies flowed from one sphere of influence to the next in a pattern that was to be repeated in the centuries to come. It was the creation of the first fire engines, however, that provided our first fighting chance against the flames.

THE BRINK OF CHANGE

During the sixteenth and seventeenth centuries, steps were being taken toward sensible fire fighting strategies, both in America and in Europe. Inventors and builders tried to get into the fire fighting game and manual pumps began to appear. Ideas passed throughout Europe, finally combining to produce an effective and worthwhile fire engine in England. Eventually, no country would be able to keep up with the innovative fire engine designs emerging from the UK. and, subsequently, its American colony. Both became major players in the field of fire apparatus very quickly. This trend defined the history of fire apparatus for centuries to come.

Change, when it came, was sometimes hailed and sometimes hated, but rarely ignored. Whether in England or America, the technology that was being born was new and exciting, but it still relied on the strength and stamina of those operating it. The secret lay in channelling human energy and employing it to greater effect. The secret lay, in fact, in ancient Alexandria, in Ctesibius' air chamber, about to make another appearance, this time on the world stage.

THE GIFT OF FIRE

And have I even given them fire.
Prometheus Bound (Aeschylus)

According to Greek mythology, Prometheus gave fire to humanity, a gift stolen from the gods. Zeus had conquered his rivals, the Titans, in battle and established himself atop Mount Olympus. Prometheus, the son of a Titan, ingratiated himself with Zeus but held a secret grudge over the destruction of his race. In revenge, Prometheus encouraged humanity whenever he could.

After being humiliated one time too many by Prometheus, an enraged Zeus took fire away from mankind, thrusting them into a cold darkness. Prometheus, refusing to give up, sought out the island of Lemnos and stole holy fire from the forges of Hephaestus, hidden in a hollow stalk. This fire was brought to humanity and mankind was able to come out from the darkness once and for all.

Zeus ordered Prometheus captured and chained to Mount Caucasus, where, each day, a great eagle sent by Zeus ate out Prometheus's liver, which grew back during the night only to be eaten again the next day. As for humanity's involvement, Zeus sent Pandora to Prometheus's brother, Epimetheus where she opened her infamous box and released sickness and evil into the world. But fire had been given to the world and we would never be the same again.

LEFT: **An oil painting of New York fire fighter John Baulch by Joseph H. Johnson, during the 1850s. The cigar is an ironic but appropriate touch.**

RIGHT: **A hand extinguisher produced by Merryweather & Sons, one of England's most brand names.**

Out of the Frying Pan

2

THE RE-EMERGENCE OF THE AIR CHAMBER AFTER

ALMOST TWO THOUSAND YEARS MARKS THE FIRST

SIGNIFICANT BURST OF GROWTH IN FIRE ENGINE

EVOLUTION. IT REAPPEARED IN GERMANY IN 1655

AND WAS LATER PATENTED BY RICHARD NEWSHAM,

THE SELF-PROCLAIMED "INVENTOR" OF THE AIR

CHAMBER MANUAL PUMP. NEWSHAM'S DESIGN

PROVIDED A BLUEPRINT FOR MANY FIRE ENGINES

TO COME AND MOTIVATED FIRE ENGINE DESIGNERS

TO PRODUCE AN EFFECTIVE FIRE ENGINE OF THEIR

OWN. THIS WAS THE AGE OF THE FIRST FIRE ENGINE

BUILDERS, DEDICATED TO THE PRODUCTION OF

THESE VITAL MACHINES.

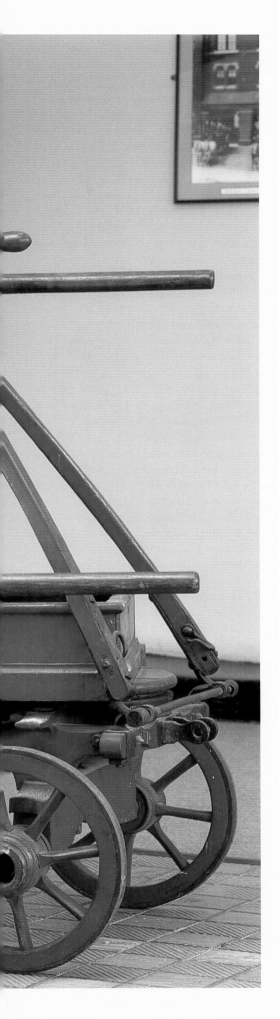

The Rise of the Manual Pump

At the end of the fifteenth century and the beginning of the sixteenth, the potential of water pumps for fighting fires still hadn't been realized. Fire had blazed and people did their best with the tools they had. Most viewed fire with resignation - sometimes it was best to stand back and let it burn itself out. In the late fifteenth century, an effort was made to use available technology to fight fires – the first recorded fire engines put into service since the Roman Empire.

Developments flowed between countries fairly quickly. If one city or country came up with a new way to fight a fire, that same idea would pop up elsewhere shortly thereafter, but the centers of early fire engine development were undeniably England and, shortly thereafter, America. The Roman Empire had left its mark on the UK and as England came to the front of the world stage, its influence was felt across Europe and, finally, in America, which in turn became a significant influence on the fire engine world. The Roman Empire had absorbed ideas from other cultures and made them its own – England and America were now set to do the same. The British would produce the first true fire engine, taking small elements from various sources. America followed the same pattern – entrepreneurial builders took the British invention and made it better.

One of the first fire engines built was a heavy brass "syringe" which needed three men to operate and was not very effective. The inventor of this ingenious but ineffectual engine is unknown but it was probably a metalsmith, since they had the skills needed to build these machines. Two men held the lengthy "squirt" between them while a third pulled back on the end of the syringe to draw in water through the nozzle, then forced the plunger forward as his comrades pointed the squirt at the fire – doing little more to fight the fire than three men with buckets could achieve.

OPPOSITE PAGE: **A side-stroke manual pump from 1839, likely a parish pump, used in England.**

ABOVE: **The Newsham engine imported to New York from England in 1731, one of the first of its kind in the United States. This model is on display at the New York's Museum of Fire.**

PREVIOUS SPREAD: **Teamwork, organization and the quality of engines were all put to the test through trials in the United States and Europe. These events were popular during both manual and steam engine development.**

These syringes were used in England and America but may have originated in Germany. Alton Plater, a goldsmith working in Germany, built a large squirt in the late sixteenth century and placed it on wheels, presumably for greater mobility. It was still syringe-shaped but much larger. Water was poured into the main cylinder, not drawn in through the nozzle. Wheels would have been helpful if the fire lay straight ahead – there were no independent axles for turning. It took a few years before anyone added a fifth wheel so the engine could be turned without having to be lifted.

Sled rails replaced wheels in some versions and a long single lever was added to allow more than one person to pump. The syringe mechanism became a piston and pumped out streams of water from a large basin filled by an active brigade of water-carrying citizens. Curved sled rails may have helped pump this single-piston system as several men operated a single lever running from the top of the engine. The momentum of the rocking engine, kept in rhythm with the pumping, may have helped the pumpers distribute more energy and produce a greater stream of water. The stream was interrupted with each upward thrust as the piston pulled in water, but the rocking action may have added speed to the operation. Another possibility is that the rails made it easier to move the engine in the mud that no doubt accumulated as water carried to the pump was spilled. This is only speculation, though. Very little is known about these machines except that most used single- pump systems and required a number of men to operate.

England was quick to adopt and improve this design, brought to the country by new arrivals such as a Huguenot refugee named De Caus, who described a similar engine in his British-published book *Forcible Movements*. The engines outlined by John Bate in his *Treatise on Art and Nature* of 1634 reveal designs that took the wheeled syringe one step further and added a large basin and levers to a single- or double-piston system similar to the model described by Hero and Pliny. These small steps brought the fire engine to its next stage of development. As London overflowed with people and new settlements were built in America, there was a growing need for a more effective means to battle a fire. A fire engine on wheels brought larger engines such as Alton Plater's to a fire, and Bate's basin and levers sprayed even more water at the scene, but there was still one problem: how to get a consistent, substantial quantity of water onto the fire? The solution was still a few years away.

FIGHTING THE GREAT FIRES

Early engines were manually operated, relying on brute strength. It was backbreaking work and strength alone was often not enough against the fires that raged through the crowded cities of Europe and colonial America. The Great Fires of Boston and London are two of the most notorious – one for the lack of preparation and the other for the preparation that did nothing to stop the damage. These fires set off an explosion of fire engine development and fire fighting techniques.

The initial Great Fire of Boston – there were two – broke out on January 14, 1653 and was the first major fire in American history. It was devastating. Boston had made no provisions for such a catastrophe and the fire spread quickly. It destroyed a third of the town but thankfully caused only three deaths. The city's main defense was a long pole with a swab at the end – Boston did not even have the benefit of a squirt. Efforts were uncoordinated as houses were torn down to create fire breaks. The fire burned itself out and left a very deep scar in the city. Fire was suddenly everyone's concern – Boston, like New Amsterdam, had had enough. Fire fighting equipment was bought and built and citizens were expected to do their part. In addition, Boston commissioned a fire engine – the first mention of a fire engine built on American soil in recorded US history. Joseph Jenks (or Jynks), an iron worker, was asked to build the engine – city records state "the selectmen have power and liberty hereby to agree with Joseph Jynks for Ingines to convey water in case of fire" – but there are no records of its appearance or effectiveness. It may not have been very good, since Boston imported an engine from England in 1678 after another "Great Fire" destroyed 45 buildings and various warehouses. The effects of the Great Fire of Boston continued to be felt throughout the colonies right up to the American Revolution.

America was not alone in its fiery tragedies. London was full to bursting and the city streets filled with more and more people each year – it was a tinderbox that required only a small spark to be set off. The Great Fire of London in 1666 was not that city's first major blaze, nor would it be the last, but the speed and extent of damage done was frightening. There had been a similar fire in 1212 – it was even called the Great Fire of London – but it paled in comparison with the ferocity of the Great Fire of 1666. The closely packed houses overshadowed the streets and wooden construction was used everywhere. Fire prevention regulations were at a minimum and only the most basic tools were available, including buckets, hooks, and ladders. Preventative measures were in place but were largely ignored. The result: fires spread through the city with ease.

Manual engines such as the squirt were available but they were so basic and so few they were only as effective as the bucket brigades doing the majority of the fighting. Houses were pulled down to create fire breaks, much to the chagrin of the owners, and many people tried to escape to the opposite side of the Thames with their salvaged property in tow. The mayor initially ignored the threat posed by the fire, then rode on horseback across the city in an attempt to organize fire fighting efforts, but few listened – people wanted to save their property and were not willing to tear down their houses until it was too late. After four days, the Great Fire of London finally burned itself out. Four people were killed and approximately 13,000 houses were destroyed.

In both cases, the effect of these Great Fires was very real and the response immediate. A fire had been lit beneath the feet of the authorities, both figuratively and literally, and the search was on for a practical, effective, and functional fire engine.

While it is tempting to think of governments making a concerted effort to promote fire engine manufacturing, the truth is the builders came to the authorities. These Great Fires, while causing a public outcry, inspired opportunistic efforts more than any generous contribution to people's overall well-being. Just as the boats gathered on the shores of the Thames charged an outrageous amount to those hoping to escape the flames, fire engine builders emerged because there was a need they hoped to supply. Necessity is the mother of invention, but it is also the step-mother of opportunity, and there were plenty of builders on both sides of the Atlantic more than willing to take on the challenge.

OPPOSITE PAGE: **The unforgettable Great Fire of London, 1666. The mayor of London rode through the city to direct fire-fighting efforts, but to little avail.**

THE FIRE OF LONDON, 1666.

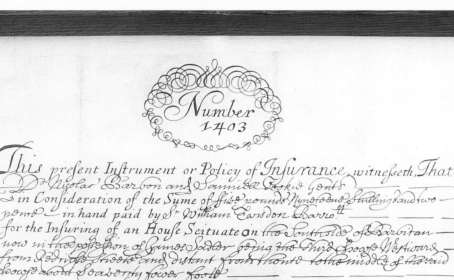

Number 1403

This present Instrument or Policy of Insurance, witnesseth, That Dr Nicolas Barbon and Samuell Tookie Gent^s in Consideration of the Sume of five pounds Nyneteene Shilling and two pence in hand paid by Sr William Earsdon Barro^{tt} for the Insuring of an House Scituate on the Southside of Barbitan now in the possession of James Sadler being the Third House Westward from Redriffe Streete and distant from thence to the middle of the said House about Seaventy fower foote

for the Term of Thirty One Yeares from the Date hereof, Do desire, direct, and appoint, That the Trustees for the time being for Houses and Lands, settled for the Insuring of Houses against Fire shall pay or satisfy unto the said Sr William Earsdon his Executors or Administrators, [or two or their Assigns, by Endorsement on this present Policy] the Sume of One hundred and thirty Pounds at the end of Two Months, after the said House shall be Burnt down, Demolished, or Damnifyed, by, or by Reason or Means of Fire; and so often as any New House, to be Built in the place thereof, shall be Burnt down, Demolished, or Damnifyed, by, or by Reason or Means of Fire, within the said Term of Thirty One Years the like Sume of One hundred and thirty Pounds If the said Dr Nicolas Barbon and Samuell Tookie and their Participants, or some, or one of them, his or their Heirs, Executors, Administrators, Agents, or Assigns, shall not within the said Two Months, pay unto the said Sr William Earsdon his Executors, or Administrators [or such his or their Assigns] the said Sume of One hundred and thirty Pounds Or in case the said House, or such New House, be only Damnifyed: Then, if such House be not Repaired, and put in so good Condition, as the same was before, at the Charge of the said Dr Nicolas Barbon and Samuell Tookie and their Participants, or some, or one of them, his or their Heirs, Executors, Administrators, Agents, or Assigns, within Two Months next, after such Damnification shall happen. Witness our Hands and Seals, the Second day of August Anno Dom 1682 Annoq^s Regni Regis Caroli 2^{di} Angl. &c. Tricesimo quarto.

Nicolas Barbon:

Sealed and Delivered
in the Presence of

Jno Bland
Sa: Edmond
Will: Dalton

Sam: Tookie

Fire Insurance Policy Issued in 1682
by Dr Nicholas Barbon's Fire Office.

DR. NICOLAS BARBON

In 1667, Dr. Nicolas Barbon, a physician who had made a profit rebuilding houses destroyed by the Great Fire and leasing them to their former owners, created the first ever fire insurance company. He offered to pay the construction costs of any customer's house destroyed by a fire. The idea spawned a flood of copycat insurance companies, such as the Hand-in-Hand Fire and Life Insurance Society, within a very short span of time.

Barbon created a fire insurance brigade in order to minimize claims, a group of fire fighters assigned to put out fires in insured houses. This team was paid well for its services and was very effective even with its simple pump engines. Insured houses were given a lead plaque or "fire mark" to display which read "Barbon's Fire Office." The brigade would hustle to a fire and try to put out the blaze – if the house was a client's. If not, these houses were sometimes left to burn or only a minimal effort was made to douse the flames. These tactics were often imitated by other fire insurance companies until some decided that putting out the fire of an uninsured house was very impressive advertising. This changed the face of fire insurance permanently. Brigades raced to be first at a fire, if only for the notoriety – some brigades put out fires in houses insured by competitors to steal the business.

It's difficult to judge Barbon harshly for what seemed outright opportunism. On the one hand, he organized a coordinated fire fighting team but on the other hand, it promoted a rivalry and pride between fire fighters.

THE NEWSHAM ENGINE

Fire "squirts" had grown into wheeled, boxed, manual pumps, but they still lacked power and consistency. It would take one more important element to complete the fire engine picture: the return of the air chamber. The debate over who was the "inventor" of the air chamber is a controversial one. Ctesibius's air chamber was innovative and an effective pump but its value as a fire fighting engine was not recognized until Richard Newsham took the stage. Newsham, an English manufacturer of pearl buttons was the first to patent the idea, on December 26, 1721, as well as the first to perfect its practical application.

There had been other engine builders who had used air chambers of various designs in their engines, but no-one had done it so effectively. Hans Hautch, for example, a German, created an air chamber pump in 1655, but its historical significance was not recognized until long after Newsham's engine had been introduced. Hautch's pump produced a consistent stream of water, like Ctebesius's model, and the nozzle had a double-jointed neck that could be pointed in any direction, but it was not hailed as a groundbreaking achievement. In 1720, another German engineer named Leupold built a small single-cylinder copper engine with an air chamber system, but it went more or less unnoticed by both the public and historians. Only Richard Newsham received the praise of other builders and inventors and he carried his banner with great pride.

The basic Newsham machine was hand-operated and contained two single-acting pump cylinders (approximately 4 inches in diameter) in a tank that made up the model's main body. It could be operated either by a pair of handles or "brakes" running the length of each side of the engine ("side-stroke" brakes), by "end-stroke" foot pedals (located at each end of the engine), or, in some models, by both together. The earliest models were plain-looking wooden boxes on metal-lined wooden wheels, hand-drawn using a wooden crossbar. The nozzle was called a gooseneck because of its distinctly gooselike shape. A team of from seven to twenty men was needed to keep the machine pumping, one of whom directed the gooseneck nozzle at the fire. A bucket chain or bucket brigade usually brought water to the engine's tank or "tub" but, for a small extra charge, some of Newsham's engines were supplied with a small double-ended leather piping suction system that could draw water both from the tank and directly from a water source.

Newsham's design had a profound and long-lasting effect. The engine's basic elements were to be found in every manual pump that was built after Newsham's was patented. Single-stroke and double-stroke engines, gooseneck nozzles, pumping brakes or handles, suction hoses and foot treadles became common knowledge to the public and fire fighters alike.

Newsham was very careful to issue instructions to purchasers of his engine on the proper maintenance of the machine – reading them today, they are as clear as your average VCR instruction manual, although you would never oil the insides of your VCR:

> Whenever the Forcers [pistons] move stiff in the barrels, pull up the Board with the Maker's Name upon it, then a little Board which lies over the Barrels, and oyl [sic] the inside of them when each Forcer is half way down; oyl also the Chains and both Ends or Pevets of the long Iron Spindle with Sallet Oyl, but grease the Elbow Screws with tallow, and do not screw them to [sic] hard on, neither screw them of [sic]; If the Chains become slack, take the Fork Key and screw the Nut a little harder, which is the Top of each Forcer, until the Chains are tight.
>
> When you lay the Leather Pipes by for some time, drain the Water well out; if then they become stiff and hard with much using, liquor them with Neats-foot Oyl, Bees-Wax and tallow, or the like, when they are a little Wet; And when they are dry'd then coil them up as at first, and hang them in a dry Place.

By Newsham's own account, his engine was the very best on the market. He may have been right, but he may also have been blowing his own horn a bit too loudly. The following description, as it appeared in an advertising circular of the time, illustrates Newsham's flair for salesmanship as well as his confidence in his invention:

> Richard Newsham, of Cloth Fair, London, engineer, makes the most substantial and convenient engine for quenching fires, which carries continual streams with great force. He hath play'd several of them before his majesty and the nobility at St. James, with so general an approbation that the largest was at the same time ordered for the use of that royal palace. The largest engine will go through a passage about three foot wide, in complete working order, without taking off or putting on anything; and may be worked with ten men in the said passage. One man can quickly and with ease move the largest size about in the compass it stands in and is to be played without rocking, upon any uneven ground, with hands and feet or feet only, which cannot be paralleled by any other sort whatsoever ... The staves that are fixed through leavers, along the sides of the engine, for the men to work by, though very light, as alternate notions with quick returns require; yet will not spring and lose time the least; but the staves of such engines as are wrought at the ends of the cistern, will spring or break if they be of such length as is necessary for a large engine when considerable power is applied; and cannot be fixed fast, because they must at all times be taken out before the engine can go through a passage ... As to the treadles on which men work with their feet, there is no method so powerful, with the like velocity or quickness, and more natural safe for the men. Great attempts have been made to exceed, but none yet could equal this sort; the fifth size of which hath played above the grasshopper upon the Royal Exchange, which is upward of fifty-five yards high, and this in the presence of many thousand spectators ... Those with suction feed themselves with water from a canal, pond, well, etc., or out of their own cisterns, by the turn of a cock, without interrupting the stream ... and play off large quantities of water to a great distance, either from the engine or a leather pipe or pipes of any length required ... The five large sizes go upon wheels, well boxed with brass, fitted to strong iron axles, and the other is to be carried like a chair.

At 60 pumps or "strokes" per minute, the smallest of Newsham's engines (there were six sizes offered) could discharge 30 gallons per minute (gpm), and the largest, 170 gpm. Newsham had created a complete and effective fire engine, sturdy enough to still be displayed in fire museums in England and the United States.

Newsham had competitors, as his promotional material implies. His greatest challenge came from an Englishman named Fowke, who built an engine with a double-spout system – according to Fowke, the first and only of its kind in the UK. In the sales material quoted above, Newsham asserted that "playing two streams at once, do neither issue a greater quantity of water, nor, is it new, or so useful, there having been of the like sort at the steel-yard, and other places, thirty or forty years, and the water being divided, the distance and force are accordingly lessened thereby." Fowke, however, claimed to have tested his machine against Newsham's and beaten him fair and square.

The popularity of Newsham's engines attracted attention overseas and arrived in New York in November, 1731. His were not the first engines imported but they were among the most sought after and durable, as Newsham claimed. The first two Newshams ordered by New York were very well received and named engines No.1 and No.2.

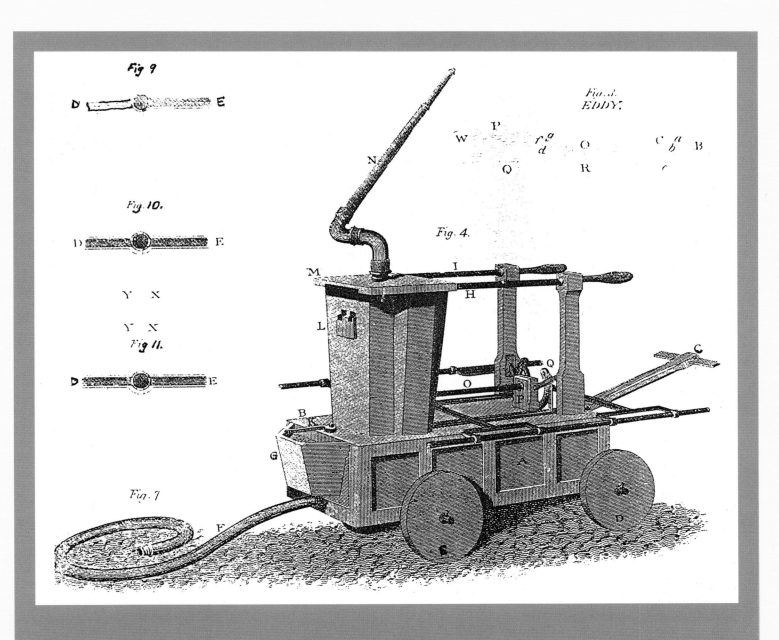

Fig 9

Fig. 10.

Fig 11.

Fig. 7

Fig. 4.

Fig. 3.
EDDY.

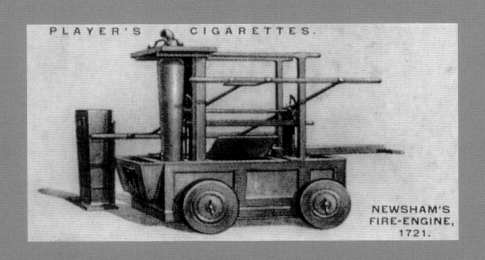

PLAYER'S CIGARETTES.

NEWSHAM'S
FIRE-ENGINE,
1721.

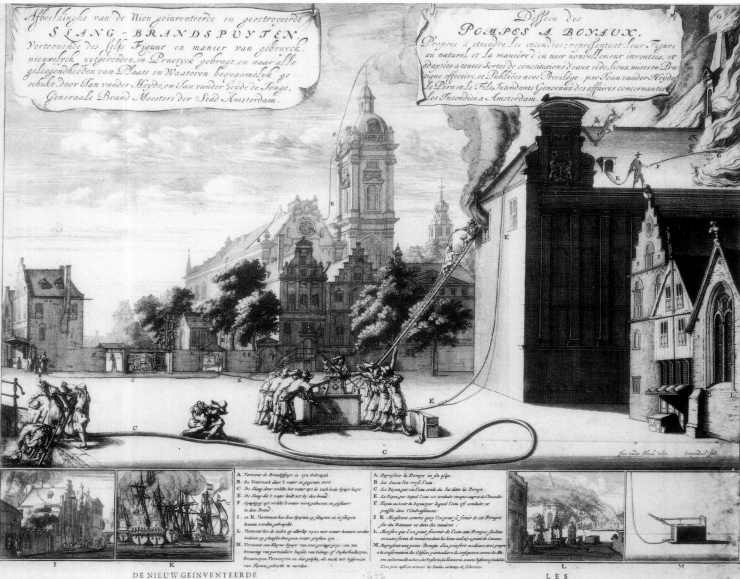

I

K

A. Vertoont de Brandtspuyt in zyn Gebruyck.
B. De Waterzack daar 't water in gegoten wort.
C. De Slang door welke het water uyt de zack in de Spuyt loopt.
E. De Slang die 't water leidt tot by den brand.
F. Spuytpyp uyt welke 't water wort gedreven en gestuurt in den Brand.
I. en K. Vertoonen hoe deze Spuyten op schuyten en in schepen konnen werden gebruyckt.
L. Vertoont hoe de zelve op allerley wyze met water konnen worden bedient, op plaatsen daar geen water grachten zyn.
M. Vertoont een Kleyne Spuyt van zeer geringe prys, om tot bewaring van particuliere huysen van belang, of Suykerbackeryen, Brouweryen, Verweryen en dier gelyke, als mede tot besproeien van Tuynen gebruykt te worden.

A. Represente la Pompe en son usage.
B. Le Sac, ou l'on verse l'eau.
C. Le Boyau, par où l'eau coule du Sac dans la Pompe.
E. Le Boyau, par lequel l'eau est conduite jusques auprès de l'Incendie.
F. Tuyau au bout du boyau, par lequel l'eau est conduite et poussée dans l'Embrasement.
I K. Monstrent comme quoy l'on peut se servir de ces Pompes sur des Bateaux et dans des navires.
L. Monstre que l'on peut fournir de l'eau aux Pompes susdites en toutes sortes de manieres dans les lieux ou il n'y a point de Canaux.
M. Represente une petite Pompe d'un prix fort mediocre, tres propre a la conservation des Edifices particuliers de consequence comme des Maisons ou l'on travaille au Sucre, des Brasseries, des Teintureries et autres Bastiments semblables. L'on peut aussi en arroser les Jardins en temps de secheresse.

L

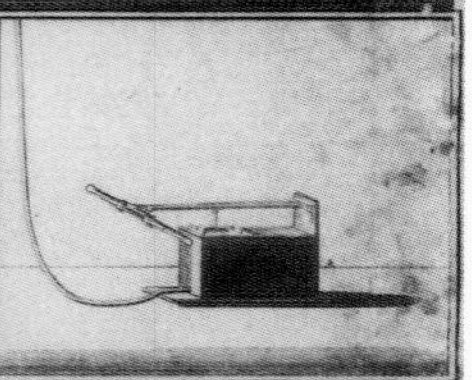

M

DE NIEUW-GEINVENTEERDE

SLANG-BRANDSPUYTEN,

Hier boven afgebeeldt, zijn dienstig en bequaam om het water met een gedurige dikke Straal te brengen in het hevigste van den Brand, en dezelve onfeilbaar en spoedig uyt te blussen hoedanig en waar die ontstaat, 't zy in nauwe straten, steegen, agterhuyzen, of in groote hoogten en andre ongenaakbre plaatzen: zonder eenige muren, wanden of gevels om ver te halen of te laten deur branden om toegang tot den Brand te krijgen: vermids daar mede het water word gedreven en geleyd door een lange en buygelijke buys, om hare gedaante SLANG genaamt, die men kan verlangen naar eys en welgevallen: en leyden zo wel krom, als regt, door deuren, vensters, glazen, gaten in muren of daken, tussen, door, en over huyzen, kerken en torens, al waar den Brand zou mogen wezen; werdende des zelfs Spuyt-pijp gebragt tot digt by het vuur; alwaar onder 't gestadig spuyten den Brand tot de laatste vonk toe wort vervolgt en uyt-gebluft. Gelijk by E en F is aangewezen.

Welk blussen zeer vaardig kan begonnen en ligtelijk onderhouden worden, als konnende deze Spuyten door 5 a 6 persoonen zonder paarden by den Brand gebragt, door 15 a 16 mannen bearbeyd, en door 5 a 6 mannen met water gevuld worden; die met gelijk getal van 5 a 6 Brand-emmers, door middel van de toebehorige zak en Slang (by B en C vertoont,) op effe straten, of anders op ongelijke en bergachtige, door middel van de Slang-Pomp en Slang (of kleyne Spuyt by M vertoont) de Brand-spuyt (schoon ver van 't water staande) gedurig met overvloed van water konnen voorzien.

En konnen meede bequamelijk op Schuyten gebruykt worden, ter plaatze daar tusschen de huyzen en 't water geen Straat is, gelijk by I; en in Scheeps Walen of Havens als by K te zien is; konnende de Slang geleyd worden waart de gelegentheyt vereyst. Konnen ook 's winters by beslooten water op 't Ys gebruykt worden, wanneer het zelve maar zoo sterk is dat het 8 of 10 menschen by malkander kan dragen.

Zijn ook bequaam om op Oorlogs of andre groote Scheepen gebruykt te worden, zo om Brand in de zelve te blussen, als om de Zeylen te natten; als mede om in ongelegentheyt van groote lekken, het water uyt het Schip te pompen, om alzo het zinken voor te komen.

De kleyne Brand-spuytjes die by M vertoont worden, zijn zeer bequaam om in particuliere huyzen mede tot blussen van Brand gebruykt te worden, en inzonderheyt om Brand in 't begin te stuyten.

En zijn deze Spuyten zedert den Jare 1672. t' Amsterdam in alle voorvallen van Brand gebruykt: en van zodanigen effect bevonden, dat d'oude Spuyten, waar mee de Stadt te vooren was voorzien, aanstonts heel buyten gebruyk zijn geraakt, en ook na rijp beraat geheel verbrooken en vernietigt, benessens alle d'andere oude Blus-gereetschappen der zelve Stadt, als Brandleeren, Haken, Zeylen en 't meerderdeel der Brand-emmers; waar tegens die wederom alleen met deeze Spuyten is voorzien; sedert welke verwisseling (die voor omtrent 4 Jaren is volvoert) men klaarlijk heeft bevonden dat de Jaarlijkse schade, die onder 't gebruyk der oude Blus-gereetschappen, voor den aanvang dezer nieuwe, door Brand wierd geleeden, meer als acht en tnegentig hondertste is verminderd. Zijnde het effect van Brand-blusschen ook tot zo grooten zekerheyt geavanceert, dat nu laatst in 26 Branden die achter een zijn gevolgt, in 't geheel niet de waarde van een eenig gemeen huys, aan huyzing en goederen is verteert; en zulks niet tegenstaande, dat veel van de zelve Branden, in d'alder felst brandende stoffen, als Peck, Teer, Swavel, Herpuys, Olie, Traan, Riet, Rijsbosse en diergelijke, ook in zeer felle Vorst, Wind, en in d'alder ongelegenste tijden van de nacht, zijn voorgevallen.

'tVoornaamste nu daar inze d'oude Spuyten overtreffen, bestaat in 't volgende:

Dat deze Spuyten veel spoediger by den Brand gebragt en gaande gemaakt konnen worden.

Datze gehanteert en bearbeyd konnen worden door veel minder volk.

Datze meer als 6 maal zo veel water effectivé in den Brand brengen.

Datze het water 2 a 3 maal zo hoog en wel 10 maal zo veer konnen brengen.

Dat deze by na noyt, en d'andre zeer dikwils ontstellen.

Datze alle Branden ten eersten en aan alle kanten konnen bereyken, hoe ongenaakbaar die voor andre Spuyten mogen wezen.

Dat de menschen die daar aan arbeyden haar in geen pericul behoeven te stellen van beschadigt of bezeert te worden, alzo de zelve veer van den Brand mogen afstaan.

Datze geen Brandleeren, Haaken, Zeylen, of groot getal Brandemmers behoeven; 't welk de kosten der zelve, die zeer groot is, uytwint; invoegen dat deze Spuyten met haar toebehooren veel minder in koop en van onderhout kosten als d'andere, en zelfs ook dat de Steeden en Plaatzen die d'oude Blus-gereetschappen hebben, zig van deze nieuwe konnen voorzien, en haar effect genieten, voor niet, of wel met voordeel, om dat, wanneer (by exempel) 1000 gulden aan d'inkoop en 't onderhoud van deze Spuyten moet worden besteet, daar tegen wel 1200 gulden of meerder kan verkregen en geprofiteert worden uyt de waarde ende kosten van 't onderhoud der oude Blusch-gereetschappen: 't welk aan yeder die 't begeert, uyt het exempel van Amsterdam en andere ligtelijk kan worden getoont.

LES

POMPES a BOYAUX.

Nouvellement inventées & representées cy-dessus, sont propres & servent à porter l'eau incessamment & à gros rayons au plus fort de l'Incendie, & à l'eteindre infailliblement & en peu de tems, de quelque maniere & en quelque endroit que le feu se soit pris, fust ce dans des Ruelles, Cus de Sac, Arriere-Maisons, ou à des Bastiments fort elevez & lieux inaccessibles, sans abatre aucune Muraille, Paroy ny Frontispice, & sans atendre que le feu y ait fait ouverture pour avoir accé à l'Embrasement, veu que l'eau y est chassée & conduite par un long Tuyau flexible, qui à cause de sa figure se peut appeller BOYAU, *lequel on peut allonger & accourcir selon le besoin ou comme on le trouve à propos, & le mener droit ou tortu par les Portes, Fenestres, Vitres, Trous de Toits ou de Murailles, par dessus, entre & à travers des Maisons, Tours & Eglises, au lieu ou l'Embrasement pourroit se trouver, de sorte que le Tuyau, qui est au bout du Boyau, est porté jusque sur le feu, où par un iet d'eau continuel l'Incendie est poursuivy & éteint jusques à la derniere étincelle: comme montrent les lettres E & F.*

Cette extinction peut se commencer promptement, & se poursuivre avec facilité, veu qu'il ne faut que cinq où six Personnes sans chevaux, pour mener cette Pompe au lieu de l'Incendie, où quinze à seize hommes suffisent pour la faire agir, & quatre ou cinq autres pour la remplir d'eau, qui quoy que la Pompe soit eloignée du Lieu où elle se puise, peuvent luy en fournir incessamment & en abondance, dans les Rues de plain pié, par le moyen du Sac & du Boyau marquez en B & C; & dans les Lieux inegaux & montagneux, par les Pompes a remplir ou petites Pompes representées en M.

L'on peut aussi en user commodément sur des Bateaux, dans les Lieux où il n'y a point de quays entre l'eau & les Logis, comme montre la Figure I, ou dans les Havres & Ports de Navires, comme en K; veu que le Boyau se peut conduire par tout où le besoin le requiert; outre qu'elles peuvent encore servir sur les eaux fermées par la Glace, pourveu qu'elle soit assez forte pour porter une Troupe de huit ou dix hommes.

Elles sont encore propres à servir sur des Vaisseaux de Guerre & autres grands Navires, tant pour en éteindre l'Embrasement, que pour mouiller les Voiles, & en pomper l'eau, quand il y en entre trop, afin de les empécher de couler a fond.

Les petites Pompes marquées en M. sont aussi tres propres pour les Maisons particulieres de consideration, & pour s'en servir lors que le Feu s'y prend; sur tout pour luy couper pié dans son commencement.

Depuis l'An 1672. L'on a usé de ces Pompes à Amsterdam en toutes rencontres d'Incendies, & l'effet en a esté trouvé si excellent; qu'aussi tost l'usage des anciennes Pompes en a esté reietté entierement; de sorte qu'aprés une meure deliberation elles ont esté toutes cassées & mises en pieces, avec les autres Instruments a etein[dre] les Incendies, dont la Ville se servoit auparavant, comme sont les Eschelles, les Cr[ocs,] les Voiles & la plus part des Seaux de cuir; au lieu de quoy l'on s'est pourveu [uni]quement des Pompes de la nouvelle invention: depuis ce changement (qui fut [fait] il y a environ quatre ans) l'on a découvert par une exacte suputatio[n ...] dommages que produisoient annuellement les Incendies, du temps qu['on em]ployoit les anciennes Pompes, & avant que l'on eust introduit l'usage de[s] nouvelles, se sont amoindris de plus de quatre-vints & dixhuit centièm[es ...] l'effet de l'extinction s'estant desormais rendu si certain & infaillible, qu[e ...] vint-six Incendies, qui se sont suivis depuis peu, il ne s'est pas consum[é ...] compris, tant en Edifices qu'en Effets, pour la valeur d'une simple Maiso[n or]dinaire, quoy que plusieurs se soient pris à des Matieres les plus combustib[les,] comme Poix, Godran, Souffre, Resine, Huile, Roseaux, Fagots & c[hoses] Semblables, méme durant les gelées les plus rudes, la vehemence des ve[nts] & l'obscurité incommode de la nuit.

Or voicy en quoi principalement les Pompes d'aujourd'hui, sur passent c[elles] du temps passé; c'est

Que ces Pompes peuvent étre menées au Lieu de l'Incendie en beaucoup moin[s de] temps, & qu'on peut les faire agir avec plus de promptitude.

Qu'elles peuvent se manier & mettre en pratique par beaucoup moins d'ho[m]mes.

Qu'elles versent dans l'Embrasement plus de six fois autant d'eau effective.

Qu'elles peuvent porter l'eau 2 où 3 fois aussi haut, & dix fois aussi loin.

Que celles cy ne se deconcertent presque jamais, comme font les autre[s] sou[vent].

Qu'elles peuvent approcher incontinent & de toutes parts quelque Incendie [que ce] soit, & quelque inaccessible qu'il puisse étre aux anciennes Pompes.

Que ceux qui y travaillent n'ont que faire de se mettre en danger d'étre offen[sez] du Feu, puis qu'ils peuvent s'en tenir eloignez.

Qu'avec ces Pompes l'on n'a besoin ny d'Echelles ny de Crocs ny de Voiles, [ny] plus que d'un grand nombre de Seaux à puiser, par où l'on en épargne la depen[se] qui est fort considerable; de sorte que pour acheter & entretenir celles-cy il en c[ouste] beaucoup moins que pour les autres; & méme que les Villes & Places, [qui] sont deja pourvuës des anciennes Machines, peuvent se pourvoir de celles [cy] pour rien, puisque par exemple s'il coute 1000 florins pour l'achat & l'entre[tien] des nouvelles Pompes, l'on retire 1200 florins, tant de la vente des ancien[nes] Machines, que de l'epargne de la depence de leur entretien; ce que [l'on] fera voir sans beaucoup de peine par l'exemple de la Ville d'Amsterdam à qui [le] que le souhaitera.

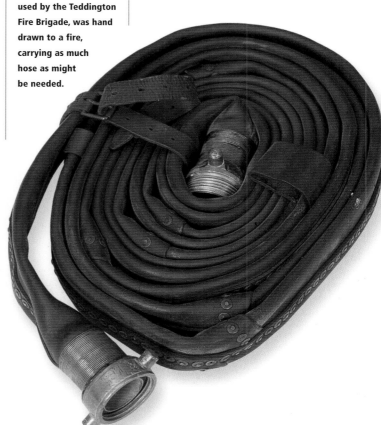

JAN VAN DER HEYDEN'S HOSE

Almost 50 years before Newsham took center stage as father of the manual pump, the Dutch inventor Jan van der Heyden created the fire hose and shaped the direction of fire engine manufacturing for centuries to come – even Newsham used leather piping on his engines. Heyden's first fire hoses were 50 feet of leather, rolled and sewn together in a single seam which usually leaked, using thread which would rot and a threaded brass screw joint to attach the hose to an engine. They would also have a substantial length of nozzle made of brass or a brass-and-copper compound, normally six or seven feet long, at the end. The unstable seam of the hose made the nozzle a necessity – hoses were known to burst under pressure. A narrow, solid nozzle would concentrate pressure at the tip and send water farther. A substantial length of metal nozzle eased the pressure on the hose itself and the risk of bursting was reduced.

As an extension between the main tank and the nozzle the hose gave "nozzlemen" more control over the water. The hose could also draw water from any nearby source and limited the need for a standard bucket brigade. If a series of engines were linked by lengths of hose, water could be pumped from quite a distance away as men worked each pump along the line.

The idea of a hose was sound but problems arose nonetheless – the sewn seam was too weak to withstand much pressure and this made firemen wary. It wasn't until James Sellers and A.L. Pennock of Philadelphia replaced the threaded seam with closely aligned copper rivets in 1808 that fire fighters put their trust in this new piece of equipment. Copper rivets made the hose more reliable, increased its lifespan, and made it ideal for fire engines. The redesigned hoses spread quickly and became a standard part of most fire engines within a few years.

Adding hose to a fire engine raised the question of transportation. At first, it was carried on a wheel or reel arrangement on top of the fire engine but this was not always successful: as firemen raced to a fire with their manual pumps, the added weight, coupled with the jostling of the pump's body, often sent the hose flying. Most engines couldn't even carry enough hose in one trip. A wagon was eventually built specifically for the hose, brought to the scene separately with as much or as little hose as was needed.

The first fire hose wagon in the United States was built by Patrick Lyon, a blacksmith and eventual fire engine manufacturer. Lyon, having moved to Philadelphia from England in 1793, was commissioned by Reuben Haines, the founder of the Philadelphia Hose Company No. 1, to produce the wagon created in 1803. His simple hose wagon, used for years, created the first official fire hose company in America.

A hose reel was set on an axle between two thin, tall wheels that could be attached to a fire engine or pulled separately around 1819. David Hubbs of New York was the inventor and prolific producer of the "jumper" as this hose wagon was known. Different sizes were built to accommodate more or less hose and these led to even more fire hose companies. Within approximately the first half of the nineteenth century, over 60 hose companies were established in New York City alone.

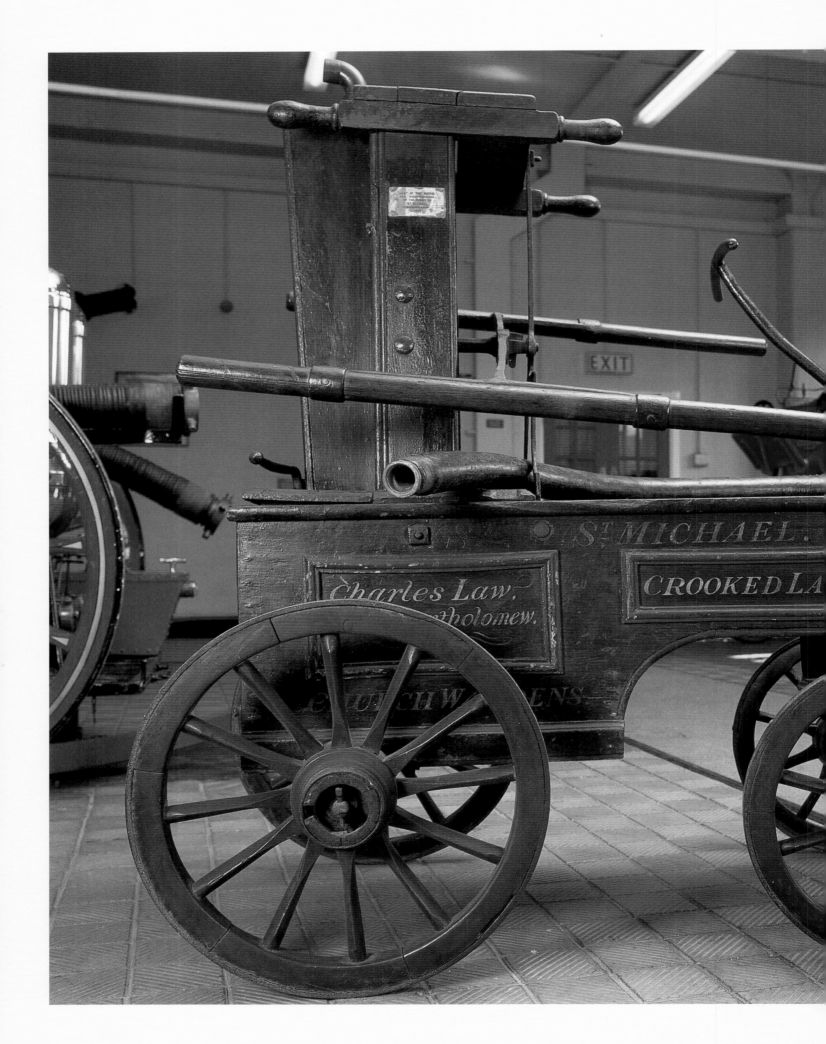

EARLY DAYS FOR THE MANUAL PUMP

Newsham's engine perfected the blueprint for manual pumpers, made possible by a solid and dependable hose. The fire engine was finally received with the respect it deserved. This early period in the history of the fire engine was marked by quick innovation and even quicker attempts to capitalize on these new ideas. The manual hand pumps produced between the mid-eighteenth century and the advent of the first steam fire engines in the mid-nineteenth century represented innovative and ingenious thinking.

On both sides of the Atlantic, Newsham's engine provided a template – his basic design could be found in most new arrivals. The only difference between American innovations and those popping up in the UK was financial: English builders were motivated largely by a competitive urge to break Newsham's early monopoly while builders in America were commissioned out of necessity.

Thomas Lote is a case in point. This boat builder created the now famous 1743 "Old Brass Backs" – so called because of the quantity of brass used as decoration – as a result of the New York City Fire Department's call for a functional fire engine. They wanted something similar to Newsham's but cheaper. Lote's was the only model that worked. Old Brass Backs was equally famous as one of the earliest practical, functional engines built and put into successful use in America, certainly the first manual hand pump built in New York. Using Newsham's design as a guide, Lote made some changes but followed Newsham's essential principles closely. He built an end-stroke engine (the pumping handles or "brakes" were placed at the ends of the engine and not along the side as in Newsham's) and moved the air chamber to the center, possibly for a smoother discharge of water. The end-stroke design may have been chosen because of the side-to-side rocking of the Newsham engine when it was in full swing.

There is always some question about so-called historical "firsts." Lote's Old Brass Backs was certainly one of the first fire engines built in America but was it the first? Anthony Nichols of Philadelphia might disagree. Philadelphia was very fire-conscious during the early 1730s and ordered two Newshams plus one more to be built by a local mechanic, the aforementioned Mr. Nichols. The engine he produced and apparently delivered in 1732 – 11 years before Thomas Lote's Old Brass Backs – is no longer around, but by all accounts it was a solid and efficient machine, similar in style to Newsham's and praised for its capabilities. Who was first is not really important. Neither one of them could have existed, after all, without the overwhelming influence of Newsham's original design.

These early years do, however, mark a rise in the independent American manufacture of fire engines and a severing of ties between the fledgling country and England. Thomas Lote's engine was slowly one such example and Anthony Nichols another. Richard Mason, a manufacturer of fire engines out of Philadelphia in the 1760s, was very successful as well. His end-stroke engine design of 1768 allowed more room for a bucket chain to keep the tank filled. More and more fire companies were springing up each year as America slowly became a country, learning how to take care of itself. Newsham's engine was the blueprint but not the end of fire engine design. It was only the beginning.

LEFT: **A manual Newsham-type pump from Bristow, England, 'owned by St. Michael parish (Crooked Lane) 1836.'**

PREVIOUS SPREAD: **Lithograph depicting two hose carts out of Philadelphia in intense competition.**

THE AMERICAN REVOLUTION

The American Revolution (1775-1783) was a significant interruption in the evolution of fire engines and fire departments but revealed close ties between American fire fighters and the political power structure. The leaders of the Revolution had fire fighting experience or connections, among them Benjamin Franklin, John Hancock, Samuel Adams and Paul Revere – even George Washington was an avid supporter of fire fighters and was eventually made an honorary member of the Friendship Fire Company of Alexandria, Virginia.

Fire fighters were exempt from service but joined the militia to fight the British in the struggle for independence nonetheless. The only problem was the particular need for capable fire fighters at that time. A perfect example is the fires that ravaged in New York after the British took the city from Washington and his fellow revolutionaries. According to Costello's history,

> Hardly had the British troops taken possession, ere (on the twenty first of September, 1776) a disastrous fire, breaking out in a small wooden house on the wharf near Whitehall, occupied by dissolute characters, spread to the northward, and consumed the entire city westward of Broadway to the very northernmost limit. In this terrible calamity, which owed its extent to the desertion of the city and the terror of the few remaining inhabitants, four hundred and ninety-three houses were destroyed …
> The cause of so many houses being burned was attributed to the military taking the directions of the fire from the firemen. The commander-in-chief, to whom complaint was made by the citizens, gave general orders that in future no military man should interfere with any fire that might happen within the city.

When the Revolution ended and the damage was assessed, it was substantial. Engines had been damaged, either sabotaged by the rebels or damaged during battles. In some cases, these early engines were stripped of their metal for munitions. The years following the Revolution were also enthusiastic as the citizens of the new United States built from the ground up. The period between the end of the Revolution and the growth of steam engines was only slightly more than half a century and yet some of the most industrious developers of fire engines appeared both in the US and the UK during these years.

THE DAWN OF THE BRAND-NAME FIRE ENGINE

Manufacturers came and went between the Revolution and the development of a steam-powered pump but each contributed something unique. In England and Europe, the Newsham design was still regarded as the forerunner of fire engine technology and manufacturers sought to make it better.

John Blanch, an English builder, produced an engine based in part on the Newsham design in 1774 that produced a stream of water 150 feet high. Charles Simpkin replaced the leather valves of Newsham's machine with metal ones during the 1790s. The wooden body of Newsham's original was eventually replaced with a metal version, its treadles and chains were discontinued, the body size was increased, and horses were added for greater speed. No matter what changes were made, however, Newsham's design still shone through the final result, a credit to his ingenuity and ability.

RIGHT: **Diligent Engine Fire Company, Philadelphia, circa 1821. The smaller illustration depicts a manufactory at work, like that of Patrick Lyons.**

DILIGENT

ENGINE FIRE C⁰

1857 by **GEO. G. HEISS**, N⁰ 213 N 2ⁿᵈ St. above Vine, 3ʳᵈ Story

Drawing taken in 1844.

Y 4ᵀᴴ 1791   INCORPORATED APRIL 1ˢᵀ 1831.

Rob.t Tempest and Jos Barton.

President & Vice President,
of the

HIBERNIA ENGINE C.o

Engraved, Printed and Colored expressly for the Firemen's Magazine by Stott & Martin.

HIBERN
CO
ins!

ABOVE & FAR RIGHT:
**Images from the
Hibernia Engine
Company, including
the uniform of choice,
a Philadelphia-style
pumper and the
company headquarters.**

LEFT: **An early
Philadelphia-style
pumper, 1797, built by
Richard Mason.**

OVERLEAF: **The
Diligent Fire Engine,
built by Patrick Lyon in
1820 and rebuilt by
James Agnew in 1835.**

Innovation and manufacturing in the United States moved at a tremendous pace. There were still ties between the US and the UK and some developments came from former British now living in the US. Patrick Lyon, the aforementioned builder of the first hose wagon in the US for the Philadelphia Hose Company No. 1, for example. Born in England, he always paid close attention to the requests and requirements of his customers. He built his first fire engine in 1794 and designed a double-decker (two levels), end-stroke, manual pump with a double-piston cylinder system and a gooseneck nozzle by 1799. The orders poured in – Lyon sold approximately 150 engines between 1794 and 1826. The Good Will Fire Company of Philadelphia was one eager buyer. In 1802, it outfitted its Lyon engine with horses, the first of its kind in recorded American history.

Lyon's competition came, in Philadelphia anyway, in the form of Richard Mason and his son Philip, who began building engines in 1768 (his son came into the business 1795). His innovative but simple end-stroke design was the smallest of the six designs Mason would

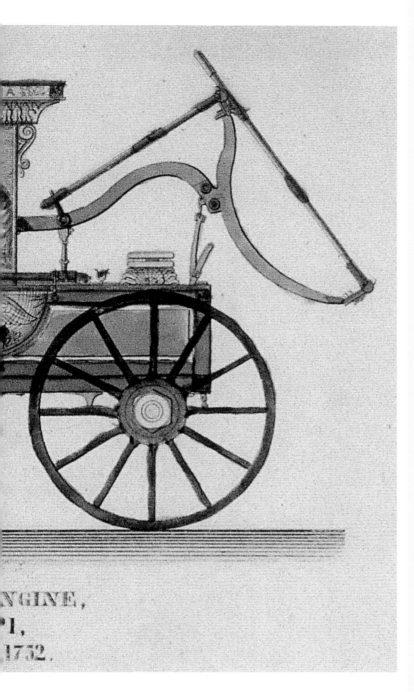

NGINE,
P1,
1752.

HIBERNIA ENGINE HOUSE,

Drawn by Hoxie & Button.

Marble Front.

eventually produce. This was a new step in fire engine design. Years later, both end-stroke and side-stroke designs would proliferate. By the time Mason's company closed down it had sold 120 engines.

One of Lyon's less successful competitors was Richard Sellers. His "Hydraulion" engine, built in the late eighteenth century, combined an end-stroke design similar to Mason's with a hose reel (a wheel set on top of the engine around which a length of hose could be wound), which was not a common feature on engines then being manufactured. Unfortunately (or fortunately for Lyon perhaps) Sellers's business did not last long, unlike that of Jacob Perkins, who was quite successful. He arrived in Philadelphia from Massachusetts in 1766 and designed his own compact engine pump in 1801. He joined forces with Thomas Jones around 1817 to form the Perkins & Jones Patent Fire Engine & Hose Pump Makers. Their largest engine could discharge up to one and a half tons of water per minute, reaching heights of 180 feet in trials and costing $1,000. Perkins claimed to have sold over 200 engines by the time the partnership dissolved in 1819.

On Stone by G.G.Heiss, 213, N.2ⁿᵈ St. above Vine.

DILIGENT

INSTITUTED JULY 4, 1791.

Heights reached by the streams of the En

Single stream to height of 196 feet 6 in. thro in nozzle.   *2 side stream to height*

Built by
Patrick Lyon, 1820.

Committee app

P.C.Eltmaker, Phœnix Hose Cº.   *S.A.Bal*

RE ENGINE

INCORPORATED APRIL 1, 1831.

rial of power over D.ʳD. Jayne's Building May 22,1852.

thro ¾ inch nozzle.   4 streams 2 side and 2 gallery 134 feet, thro ½ nozzle.

Rebuilt by
Jnᵒ Agnew, 1856.

e C.ᵒ to decide.

re H. & L.   E. Sratton, Harmony Fire C.ᵒ

RIGHT: **This lithograph from 1855 depicts a Philadelphia-style, end-stroke hand pumper and fireman from the Waccacoe Fire Company.**

OPPOSITE PAGE: 'What Boys may expect when they get in Firemen's way.' This illustration, entitled 'At a Fire', was originally published in 1858.

BELOW: A 'piano-type' manual pump, circa 1880, as popularized in New York by the Button Fire Engine Company.

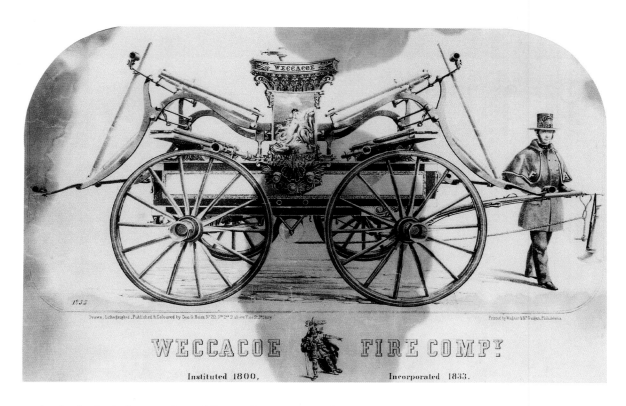

WECCACOE FIRE COMP?

Instituted 1800,   Incorporated 1833.

Philadelphia quickly became a leader in the industry – the public referred to Lyon's large end-stroke engines as the "Philadelphia-style." This familiarity with engines and engine types reflected the industry's growth and public appreciation of these machines in a growing society. Similarly, the engines of Jacob Broome and Abel Hardenbrooke, both of 1780s New York, came to be referred to as the "New York-style." These lugubrious machines were still based in part on the typical Newsham design but had come into their own with America's new-found independence. Broome's was an end-stroke system worked by eight men and Hardenbrooke's creation was a side-stroke engine with rear wheels larger than the front.

This popularity mirrored the widespread professional and public acceptance the engines enjoyed. James Smith's "gooseneck" engines with the distinctive gooseneck nozzle connections sprouting from the top of his machines became quite well known. These were built in New York City between 1810 and 1864. Goosenecks had been around for many years but Smith's was the first manufactured on a large scale – he sold about 500 during his company's lifetime. Smith also added a fifth wheel under the chassis which made turning the machine much easier – without the fifth wheel or a movable axle, engines had to be stopped at a corner and picked up by the volunteers in order to be brought around. Smith's company had the additional distinction of creating the critically acclaimed but oddly named "Shanghai" engines. Production on these engines began in the 1850s and they were celebrated for their sturdy construction. What made these monstrous engines unique was their two sets of pumping handles which operated alternately – a "ground crew" pumped in unison with an "upper deck" crew. Smith's double-deck end-stroke machines were exceptionally effective and are respected to this day. The affectionate nickname given to the engines – "Shanghai" – again reflected how popular these engines were to the public and how warmly they were viewed.

The Button Fire Engine Company, established by Lysander and Theodore Button in New York in 1834, produced the "Piano Box" style, derived from the sheer size of the water tanks used on Button engines built after 1838. Side-stroke engines with flat decks and heavy water boxes, Button engines bore only a superficial resemblance to a piano but the name stuck. Button engines also pioneered angled valves for a more level and even distribution of water, introduced adjustable leverage in 1842 to increase or decrease the amount of water being distributed (without interrupting the pumpers), and, most notably of all perhaps, created the first permanent suction hose attachment. This design was dubbed the "Squirrel Tail," because the hose was connected to a long brass tube on the deck of the engine body and wrapped up from underneath the rear of the engine and across the top, like the tail of a squirrel. The father-and-son team of the Button Fire Engine Company were among the most innovative of their day, building approximately 500 engines of seven different sizes in total.

Public familiarity with the fire engine industry sometimes had unfortunate overtones. Henry Waterman of New York, for example, was commissioned to build a double-deck engine for the Empire Engine Co. No. 42. The result was a massive, powerful side-stroke engine built during the 1840s, quickly named the "Haywagon" by volunteers. Its size made early runs a hazard – if the strain of pulling the Haywagon to the scene of a fire didn't kill you then its unwieldiness certainly could. The public renamed it "Mankiller." It was considered dangerous but never hurt anyone. The volunteers had the good sense not to use it except in the most dire circumstances.

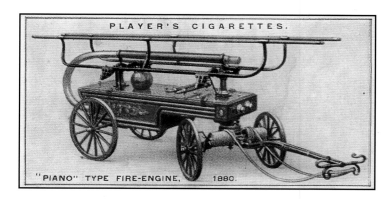

PLAYER'S CIGARETTES.

"PIANO" TYPE FIRE-ENGINE,   1880.

Entered according to Act of Congress, in the year 1858, by Henry G. Harrison & Wm. N. Weightman, in the Clerks Office of the District Court of the Eastern District of Penna.

At a Fire.

What Boys may expect when they get in Firemen's way.

Pub. by Harrison & Weightman 118 N. 10th St. Phila.

PREVIOUS SPREAD: **A restored New York-style James Smith manual pumper – the 'Deluge' – built in 1839.**

RIGHT & OPPOSITE: **Illustration of two Philadelphia fire engine companies, with prominent display of manual pumps.**

BELOW: **A hand pumper from the Hibernia Fire Company, 1836, as built by John Agnew.**

Fame was not limited to a popular engine. William Hunneman, a coppersmith from Boston, Massachusetts, is a famous engine builder of the late eighteenth century in America because he was apprenticed to Paul Revere. He began manufacturing light and maneuverable engines in the 1790s which featured pumping handles set as though for side-stroke use but which could be brought around for end-stroke use at the scene of the fire. The simple modification opened the door for much longer pumping handles and more men could pump at once. (In end-stroke engines with levers located at the front and rear ends, the length of the levers affected engine width; in a side-stroke engine, the brakes ran the length of the engine, sometimes longer if the handles could fold.) Hunneman also incorporated a "crane-neck" design on the main chassis of his engines. The two wheels at the front of the engine's main body were secured to metal shafts that bent upward and curved down, like a crane's neck. This innovation, though not entirely Hunneman's, was an outstanding selling point of his machines because it made turning easier and increased maneuverability. The success of his design led Hunneman from Boston to larger headquarters in Roxbury and enticed his sons Samuel and William Jr. to join the company. During its 90- year history, the company produced more fire engines than any other manufacturer and was one of the few to make a successful shift from manual pumpers to steam engines.

Public awareness of the average fire engine builder's name had grown beyond simple popularity. Manufacturers' names had become brand names. People spoke of Hunneman, Lyon, and Button not in reference to the individuals but their machines. These builders were not alone. There were an endless array of builders and

inventors who came onto the scene and caught the public's attention. John Cooper of New York, for example, built a rotary manual engine based on a simple rotating piston principle in 1827 that proved more powerful than standard manual pumpers of the time. John Agnew of Philadelphia produced an enormous double-decker engine in 1841 that was so large it required two sets of men to operate each handle, one on each level. Dudley Farnum and Franklin Ransom, two New Yorkers of the late 1840s, created a fascinating machine based on a rowing motion. Fire fighters sat on top of the engine and held on to a length of wooden handle attached to a central shaft which was itself attached to the central piston arrangement. The volunteers were arranged along the top of the engine on opposite sides of the main piston complex. The two teams would row in unison and created a consistent pumping system. According to the inventors' patent application, rowing let the men work for much longer with greater power. Unfortunately, only a few engines are known to have been sold.

The rise of the manual pump in the United States in particular had as much to do with public approval as with an engine's effectiveness. A new machine bought or built created quite a stir in most communities and this attention maintained the fire fighters' status and position. As the public clamored for more impressive machines, the fire fighters tried to get the most out of the attention. The result was an influx of machines and the birth of the industry on American soil, coupled with the raised stature of the fire fighter in the public domain.

RIGHT: **'Fire in London'** – this engraving dates form 1808. Note the fire fighters able to stretch their hoses right to the door of the burning building (bottom right hand corner).

BELOW: **A 'factory pattern' manual pumper, 1885, built by Merryweather & Sons for factory fire brigades.**

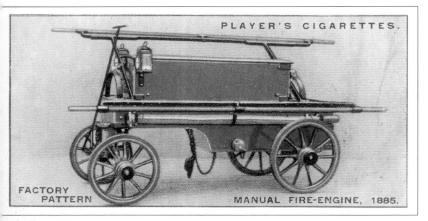

THE MANUAL PUMP IN ENGLAND

As American invention and enthusiasm for these magnificent machines grew, there was as much drive overseas to build a better fire engine. Newsham's engine had sparked fierce competition and companies tried to take Newsham's design to a new level. From the earliest manual pump developers, John Blanch and Charles Simpkin, a long line of well-known and respected manufacturers emerged, proving their worth as much as those in the US.

Simpkin had replaced the leather valves of the Newsham model with metal ones at the beginning of his career. He joined another manufacturer to create Hadley and Simpkin in England in the early 1820s and introduced foldable brakes (these created a more compact and maneuverable model like Hunneman's model with the swivelling brakes) and springs to support the engine body – effectively the first shock absorbers. William Baddelley introduced a portable canvas cistern or bag to collect grit from water drawn off a water plug – any grit in the works caused great damage in a short period owing to any increased friction.

Moses Merryweather was one of the most famous English manufacturers of fire engines at the time. He established the Merryweather Company within the former Hadley, Simpkin, and Lott firm after marrying Henry Lott's niece in 1836 and took his place at the forefront of manufacturers in England, if only because there was so little competition. The very light horse-drawn model – the "Paxton" – earned a popular reputation in more rural communities, where its lightness and maneuverability were great assets. The Paxton could reach 100 gpm and a height of 120 feet under ideal conditions. The Merryweather "London Brigade" was drawn by two horses and incorporated a strong handbrake to slow it down on slopes. The engine was mounted on springs similar to Simpkin's design and the firemen rode on top. To see this tremendous machine come hurtling down the road, drawn by enthusiastic horses and topped with volunteers must have stopped quite a few people in their tracks.

LEFT: **An early Merryweather & Sons parish pump, hand-drawn, end-stroke manual pump design, from 1758. It was presented to Mortlake Parish by Princess Amelia.**

BELOW: **Long metal pipes were thrust into London's sewers from the road above, channelling the limited water pressure through the hoses. Sewers were initially hollowed-out tree trunks running close beneath the road surface – hence the expression, 'trunk road.'**

Merryweather's notoriety was not limited to England. His company produced the St. Petersburg floating fire engine in 1835, commissioned by the Russian government and designed by Tilley, an engineer from England. This small, metal engine was heavy but effective. Merryweather also produced small, portable "Factory" engines for warehouse and prison use. These small engines were a staple in the community for a number of years. William Roberts, an English designer of the late nineteenth century, fell in line with this idea and produced a small manual pump that could be disconnected from a carriage and used, for example, on board a ship or in other compact situations.

THE RISING STEAM

The manual pump served the public interest well, even after the advent of the steam engine – in some cases, even after the invention of the motorized fire engine. Newshams, Merryweathers, and Buttons can still be seen on display today, some even put to use during special celebrations. These engines were meant to sit behind velvet ropes or in glass cages; they were meant to be used, in keeping with their original instructions, "to exercise the engines, so as to keep them in good order." It is hoped that this tradition will be continued for many generations to come.

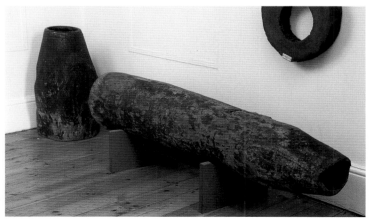

RIGHT: **Benjamin Franklin, one of the United States most-celebrated founding fathers and pioneer of the American fire insurance system.**

OPPOSITE: **In this well-known print, the tools, the uniform, the manual pump and the enthusiasm of the volunteer are well-represented.**

BELOW: **A Phoenix Hose, Hook & Ladder Company certificate from 1878.**

THE VOLUNTEERS

No history of fire apparatus is complete without mention of the early volunteer brigades. Later fire brigades in the United States and England had their own special place in history and were vital to the communities they served, but the volunteers had a special role to play. Their sweat and effort brought the earliest manual pumps to the scene of a fire and produced results. Fire brigades today are filled with technological advancements and require a fire fighter to be an expert in the field – the volunteer threw his heart into every situation. This had a profound effect on the public perception of fire fighting and forced the direction of fire engine design for many years.

Fire insurance brigades in England gave way to volunteer brigades due to the violent competition that built up between different companies – the public had grown tired of fire scenes turning into battlegrounds. Nevertheless, the rough spirit that characterized rival insurance brigades continued into the middle of the nineteenth century and ended with the advent of fully paid fire

fighting departments. Competition between fire insurance companies reached a peak as they tried to steal each other's business. Some would send runners to the scene of a fire to monopolize a fire plug or other water source. Others would start fights with another company before fighting the fire itself. Fire insurance brigades had overstayed their welcome.

In addition, not everyone could afford fire insurance. The government established volunteer brigades to protect the public as a whole. Unfortunately, the volunteers eventually became a problem themselves and caused the same stir the insurance companies had created. Competition was so fierce among volunteers that, in England at least, teams of pumpers encountering other teams dragging their enormous engines to a fire would race all the way to the scene. Egos and pride were at stake and the volunteers did not take the challenge lightly.

The US was to have its own volunteers, the first founded by none other than Benjamin Franklin, newspaper editor, a "Father of the Revolution," one of the founding fathers of the United States of America and the discoverer of electricity. Like Dr. Barbon in England before him, Franklin established a fire insurance company. Unlike Dr. Barbon, Franklin had the financial wisdom to create a volunteer fire brigade before setting up his insurance company, which meant his company did not have to support its own brigade. He contributed to the volunteer fire department directly instead.

As editor of the Pennsylvania Gazette, Franklin had called for an improvement in fire fighting techniques and fire prevention strategies. Franklin founded the Philadelphia Contributionship for the Insurance of Houses from Loss by Fire (later renamed the Hand-in-Hand Company) in 1752 and established himself as a pioneer in American fire fighting thinking. He would remain a patron throughout his lifetime.

AT A FIRE.

STORAGE ROOM

SAVING GOODS

Franklin was a supporter of fire prevention from his very early days in Philadelphia, writing about the need for greater fire safety for the Pennsylvania Gazette. It is ironic that this stalwart promoter of fire safety established himself in Philadelphia, a city that had suffered relatively few fires. A fire in 1730 provoked the purchase of a mass of buckets and three engines (two Newshams and one engine from Anthony Nichols, as mentioned earlier) but it also motivated the creation, by Franklin and four associates, of the first official volunteer fire brigade in Philadelphia. The claim that his was the first ever in America is not certain but his popularity made him the perfect father figure for early fire fighting volunteers. Franklin's brigade was founded in 1736 and called the Union Fire Company. It was independent of any governmental funding or control, made up of members only, 30 in total. By Franklin's own account, the volunteer fire brigade was so popular that it spawned the creation of others and within a very short period of time most "men of property" were involved with such a brigade in some capacity. Membership carried status.

It wasn't long before Franklin founded the Philadelphia Contributionship, the first such insurance company in America. It was similar to Dr. Barbon's in England, but relied on the volunteer brigades and supported them with donations. Franklin's model exerted greater restraint on the men involved, something that was essential at the time.

Life for a volunteer was not easy. Their experience affected fire engine development directly. They dragged their pumps to a fire scene, the pumps weighing up to 1,000 pounds, and cleared a space through the gathered crowd, often close enough to singe hair. Bystanders were handed buckets to bring water while the volunteers pumped away at the levers until exhaustion set in and a new volunteer stepped in. Early insurance brigades in England were given buckets of beer to pass around for greater resilience against fatigue! This bolstered enthusiasm but it often created a drunken stir and led to occasional brawls or cursing between competing brigades.

PREVIOUS SPREAD: **A black and white engraving of "The New York Fire Insurance Patrol" showing all aspects of a fire fighter's duties.**

ABOVE: **This 1854 lithograph of 'The Life of a Fireman' shows a heavy end-stroke engine drawn by a dozen men on the way to a fire.**

RIGHT: **Hose reels and end-stroke manual pumps are pulled in a race between volunteers in this print by Nathaniel Courier, from 1854.**

For all the good that was done, volunteer fire brigades sometimes took on the appearance of a gang, with their own rules and significant political power. Their opinion was vital to the general populace – they held public safety in their hands and this gave them serious influence. In their role as protectors of the public safety, some pumpers began to believe their own pomp. Something had to be done and quickly.

Rules were set down to limit the reach of these brigades and hold back the rough-housing. English firemen were subjected to heavy fines for any questionable actions, including swearing, challenging another fireman to a fight, or throwing water. New York's Common Council established regulations to exempt fire department volunteers from civil service such as jury duty, military service, and constabulary responsibilities as long as they agreed to be "diligent, industrious, and vigilant" in their duties, kept their engines in good working order and were always available for fire fighting duties. If a fireman was not present at a fire without good cause, he would be fined 12 shillings, (equivalent to 60 pence today) a fair sum in those days. The District of Columbia ordered fire fighting companies to be cautious in their selection of volunteers; liquor was prohibited at the scene of a fire; any troublesome volunteers were to be let go immediately, no matter how vital, and no company could claim a water source unless present and equipped at the source itself. These rules were essential to keep the peace among fire fighters, even though they were not always successful. Rising steam engine popularity made it obvious that maintaining volunteer fire departments was less effective than paying fire fighters to do the job. During these early days, however, pomp and circumstance was the order of the day. Brightly colored engines were paraded and displayed for the public on a regular basis and the volunteers were celebrated for their bravery and courage.

PREVIOUS SPREAD: **A hand-colored lithograph from "The Life of a Fireman" series, printed by Nathaniel Currier, 1854.**

ABOVE: **'The Late Firemen's Third Annual Parade in the City of New York,' from 1853.**

RIGHT: **Another print from the 'Life of a Fireman' series by Courier, 1854. All the latest in fire fighting technology is on display.**

FIREMAN.

...an your rope"

...2 NASSAU STREET

ANDREW TORNING

Volunteer fire fighting was a worldwide phenomenon, tracing its roots back to ancient Rome, finding a home in England and the United States, stretching as far as Australia. Andrew Torning, an actor, artist, and decorator born in England in 1814, arrived in Sydney, Australia, with his wife in 1842, where he worked at the 'Vic' theatre and organized its fire brigade. This was his first taste of fire fighting and it was to stay with him throughout his life. Through his efforts, the theatre fire brigade found its own home and was established as a volunteer service eventually recognized by the community.

He went to San Francisco in 1858, where he and his family lived for almost 10 years and where Andrew was once again involved in fire brigade activities. When he returned to Sydney in 1867, he was re-elected Superintendent of the Australian Fire Company No. 1 and his ties with fire fighting were confirmed. During his tenure he fought against the upsurge of insurance brigades, never trusting their motives, and sought for general fire protection for all. On his death in 1900, he was celebrated as folk hero of his day and acknowledged as the father of volunteer fire brigades in Sydney. Though few are familiar with his name today, his achievements remain for all to see and his influence is keenly felt.

3 Fighting Fire with Fire

STEAM POWER SET THE INDUSTRIAL REVOLUTION UNDER WAY. LITTLE DID INNOVATORS THOMAS SAVERY, THOMAS NEWCOMEN, AND JAMES WATT KNOW THAT THEIR WORK WOULD HAVE SUCH FAR-REACHING EFFECTS. FIRE WOULD LITERALLY BE USED TO FIGHT FIRE AS STEAM GENERATED BY BURNING WOOD AND COAL POWERED THE ENGINES THAT ATTACKED THE FLAMES. THE PRACTICAL STEAM ENGINE PRODUCED AN EXPLOSION OF IDEAS, ENTHUSIASM, AND SERIOUS COMPETITION AMONG FIRE FIGHTERS AND INVENTORS ALIKE. AN ERA OF BUCKET CHAINS, MANUAL PUMPS, AND QUESTIONABLE FIRE FIGHTING METHODS WAS COMING TO AN END, REPLACED BY A NEW, PROFESSIONAL ATTITUDE.

LEFT: **Steam replaces muscle – almost – in this Currier & Ives lithograph from 1861. The perfect example of the crossover between manual pumps and the growing steam.**

PREVIOUS SPREAD: **The steam engine, applied to fire fighting, proved a powerful ally in the ongoing battle against the flame – once the invention was accepted.**

The Great Steam Debate

Everyone was aware of the potential locked into fire, from Hero's earliest attempts to build an effective engine around the first century AD to the experiments of Thomas Savery and Thomas Newcomen, pioneers in steam technology. Savery patented a steam-powered water pump in 1698, although it was supposedly not very effective. Newcomen produced an effective and efficient pump in 1711 in which a steam-powered piston operated a crankshaft to pump water from mines. It would take a few years before anyone would apply steam power to locomotion and even longer before it was applied to fire engines, but it was only a matter of time.

James Watt's 1769 creation of the first practical steam engine was the sounding bell of the Industrial Revolution and introduced mass transport to the world. Watt's engine, based in part on Newcomen's idea, was once again used for pumping water from mine shafts. It was not considered for fire engines for a number of years but, when it was, it represented powerful and irrevocable change.

Despite the early steam engine work conducted in the UK, the idea of steam-powered fire engines did not catch on in England until well into the nineteenth century. Robert Fulton had already created a steam-powered ship called the *Clermont* in America by 1807 and made the 150-mile trip from New York City to Albany 20 years before the first practical fire engine was built. Steam-powered ships had crossed the Atlantic by 1819 but the steam fire engine remained a decade away.

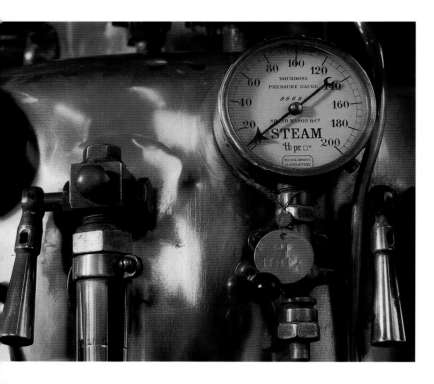

AN UNCERTAIN SWITCH TO STEAM

The clearest demonstration of English reluctance to accept the steam-driven fire engine lies in the history of John Braithwaite and his assistant John Ericsson, both working out of London, builders of the first steam fire engine. The "Novelty," as it was known, was based on a boiler stoked with burning coal in the rear of the engine connected to two horizontal cylinders running through the water fed to the pistons. The resultant steam moved the pistons connected to the pumps in the vertical air chamber near the driver's seat. Fully heated, the engine was immensely strong. Its two cylinders could reach 10 horsepower and discharged at least 150 gallons per minute (gpm) – 250 gpm at its peak – to hit reported heights of 90 feet.

Early illustrations of this machine show the boiler and pump housings at opposite ends of the body with intake and discharge hoses attached to the bottom of the pump, itself located just behind the driver's seat. Braithwaite hoped for the approval of British insurance companies which would guarantee acceptance by the London fire brigade system – a necessary tactic if the steam-powered fire engines were to come into mainstream use. Unfortunately, British insurance companies did not agree and would not endorse the machine. Complaints ran the gamut: it takes too long to get up to steam (it did take 20 minutes to power up, but could operate for much longer and with far greater consistency than any manual engine); it discharges too much water with too much power, causing more damage than good (while these machines did spread more water over a greater area than manual pumps, there is nothing to suggest that they caused more damage); it requires more water to operate than can be supplied (this was true – the limited water supply flowing through the simple sewer systems was used up very quickly by one of these engines); it is too heavy to pull to the scene of a fire!

The steam engine's worth was proven without a doubt, however, at the Argyll Opera House fire in Soho, London, during the winter of 1830. Hand pumps froze quickly while the steam pump worked for five hours without interruption and used only three bushels of coke in the process. Braithwaite and Ericsson's Novelty was used again in a later brewery fire as well as the Houses of Parliament fires of 1834 but, remarkably, the public and the British insurance companies remained unimpressed. The steam-powered fire engine would not be used officially in London until 1858, although Braithwaite made a few smaller engines, including one called the "Comet" for the King of Prussia in 1832 capable of approximately 300 gpm. Another was sent to Liverpool where it was kept in service for a number of years. The public distaste for steamers and the lack of financial backing, however, eventually forced Braithwaite to gave up on fire engines and concentrate instead on locomotives.

Ericsson moved to the United States where he continued working with steam power, building steam fire engines with some degree of success. One of his earliest designs was submitted to a contest sponsored by the Mechanics Institute of New York, seeking the best steam-powered fire engine. Ericsson's model, based in large part on the engines he built with Braithwaite, won the Gold for its tremendous power and light weight.

ABOVE LEFT: **The steam gauge –
this one from the Shand, Mason & Co. –
came to symbolize the new steam-
powered engine of the day.**

ABOVE: **A clipping depicts the typical
horse-drawn steamer in use in 1866.
So called 'self-propelled' steamers
would more often than not require
the help of horses.**

LEFT: **The infamous Ericsson
& Braithwaite steam engine.
Its stream of water could reach 90
feet. Though developed in the UK,
the steamer was not adopted for use
in London for years.**

RIGHT: The 'Hodge' steamer of 1841 – said to be among the first of its kind built in America. Note the wheels which, when raised off the ground, acted as flywheels when in operation.

OPPOSITE PAGE: The life of the fire-fighting volunteer was not easy, as suggested by these rough and ready characters of the Hand-in-Hand Fire Company No. 1 (Philadelphia), circa 1863.

PLAYER'S CIGARETTES.

THE "HODGE" STEAM FIRE-ENGINE, 1841.

THE STEAMER HITS NEW YORK

One of Ericsson's competitors for the Mechanics Institute award was Paul Hodge, a former draftsman and designer of engines in New York. Hodge was originally from England and had worked as a draftsman for an early locomotive builder before finding his true love in steam-powered fire engines. He published a book on the subject, The Steam Engine, in 1840. Hodge's interest was generated in part by the many fires suffered by New York around 1835–36, fires of such magnitude and destruction that the city's mayor and Common Council demanded action. The results are reflected in Costello's Our Fireman:

> Mayor Lawrence, in September, 1835, called the attention of the Common Council to the frequency of fires, and particularly the one in Fulton, Ann, and Nassau Street, and also the fire in Walter Street and Maiden Lane, by which a large amount of property was destroyed and lives lost … Not more than a fortnight [later] another large fire broke out in Franklin and Chapel (now College Place) Streets … Then followed the terrible conflagration of December 16, which destroyed in one night twenty million dollars' worth of property, and dislodged more than six hundred mercantile firms. By that calamity the extensive resources and irrepressible energies of the citizens were developed …

The largest of the fires mentioned above, the so-called Great Fire of New York of December 16, made it clear that goosenecks and manual pumps could not handle major fires. More versatile, powerful engines were needed and Hodge tried to fill the niche. He demonstrated his first fire apparatus in front of the City Hall in 1841, called the "Exterminator" – the first technically "self-propelled" fire engine, even if it was too heavy to move without the aid of horses. Its large back wheels acted as fly wheels for the pistons and the whole could be hand-drawn or horse-drawn. The "self-propelling" steam engine would help move the machine but with only limited success.

Unfortunately for Hodge, no-one was interested. The Pearl Hose Co. No. 28 in New York had nothing but complaints about the model they ordered and it was promptly returned. Hodge returned to England in 1847, dissatisfied with the results but enthusiastic as ever about steam's potential. Hodge was up against the power of the volunteers. New York City's volunteer brigades, though respected, held local politicians in their pockets and preferred the status quo. The situation is spelled out clearly in Costello's book:

> The element of rowdyism in the Fire Department … manifested itself at a fire which occurred on the night of January 1, 1836, when Alderman Purdy, representing the tenth Ward, was set upon and mercilessly beaten by members of Engine Company No. 10, who were also accused of abandoning their engine on that occasion. For the latter offence, nine officers and members were expelled the Department, and ten were suspended for not complying with the requisitions to appear before a committee of the Common Council and testify in reference to the assault on Alderman Purdy.

Flouting the law and abandoning the dignity that had grown with the New York City Fire Department, the volunteers were a symptom of the trouble that lay ahead.

THE END OF THE VOLUNTEERS

The need to change the volunteer brigade system was on the rise around the middle of the nineteenth century in America. Volunteers were becoming unruly and disrespectful and their attitude was becoming incorrigible but their political influence remained high. Some cities, like Cincinnati, proposed a paid fire department. The volunteers responded with predictable disgust. The proposal raised the issue of cost and it was obvious that a paid system could not support the number of volunteers that filled fire brigade ranks. Steam engines were suddenly much more appealing.

Alexander Latta, an inventor, and Abel Shawk, a mechanic, were two entrepreneurial-minded men working out of Cincinnati who recognized an opportunity when they saw it. Combining their individual expertise, they built a steam- powered fire engine which included a copper coil in the engine's boiler to speed its start-up rate, a serious concern for any fire engine manufacturer. The city agreed to order a Latta & Shawk engine and a motion for a paid fire brigade was passed. This did not please the volunteer fire brigades and Latta and Shawk had to work in relative secrecy. According to a reporter from the Chicago Herald, Latta was so concerned he feared being killed before the machine was put into service.

BELOW: **A.B. Latta's self-propelled steamer , patent 1854. Latta's engineering design spawned the creation of a string of steamers.**

The engine, named the "Joseph Ross" after its sponsor, was later nicknamed the "Uncle Joe Ross." It was almost self-propelled, requiring horses while getting up to steam or going up hills. In a public test against a Hunneman manual pumper from the Union Fire Company, the steamer pumped merrily along after the volunteers had tired. The results of the trial differ depending on the source. Some claim the Uncle Joe hit a vertical height of 240 feet while others maintain it only reached 225 feet.

Latta built several engines with his brothers in a new company called A.B. & E. Latta. These were sold to various fire departments and tested against some of the best in the land (one went up against the very powerful Haywagon in 1855 and, while it couldn't perform much better, it outlasted the Haywagon with ease). Latta received a patent for a three-wheeled self-propelled apparatus with a rear fly wheel in 1855. Shawk formed the Young America Works and built a horse-drawn steamer called the "Missouri" in 1855 – it was drawn by four horses and could hit 120 feet vertically and over 170 feet horizontally.

VOLUNTEERS IN AUSTRALIA

Trouble and change followed the volunteers worldwide, as the original UK system was brought to new territories. In Australia, for example, the volunteer brigades faced as many challenges as their cousins in England and the United States. Volunteer brigades began in Melbourne during the 1850s and protected the cities and towns until they were assimilated by the newly created Metropolitan Fire Brigade in 1891. Appliances were initially imported from England until workshops were established to build engines unique to Australia.

The following excerpt from the Melbourne Metropolitan Fire Brigade's 'Centenary' Brochure explains the complications faced by volunteers and the public alike in their efforts to fight fire:

It's a long call from the early days of Fire Brigades in Melbourne with its water carts, hand hose reels, "The Old Iron Horses," as they were called, drawn to the fire by firemen, fifteen or twenty pulling with long ropes attached to the heavy cumbersome apparatus; hose carts, manual engines, and the clanging of fire bells to the well trained firemen of to-day well equipped with the latest appliances.

By way of introduction a short outline of the changes and practices that led up to the formation of the present Brigade ought to be interesting. In the very early days on Eastern Hill opposite to the present head quarters, where the Eye and Ear Hospital now stands, there was a huge tank of water conservation. Here, water carts, large barrels on wheels (often the cause of serious accidents) were filled every night ready to rush to a fire, a reward being offered for the first and second cart arriving to fill the canvas tank to supply the manual engine.

Reticulation of Melbourne provided for the fire plug, hydrant and direct water pressure, which, for many years, however, was inefficient and unreliable. It was with this poor service the Volunteer Fire Brigades, with the twenty men of the Insurance Brigade (Superintendent Hoad in charge), had for years protected Melbourne from the ravages of fire.

There were two classes of Volunteer Companies, the Municipal, and the purely Volunteers; also three or four private Brigades, such as the Carleton Brewery and Albion (Timber Yard). Fire bells would give the alarm, and dozens of Brigades with hose carts, manual engine sand hand reels, with large ropes attached, drawn by firemen who often ran for miles to get to the fire. The idea was "First Water," failing that, get there and get to work, to ensure the name of your brigade being in the papers.

Every Brigade did that which seemed best in their own judgment. As a result, there was often disorder and chaos. Thoughtful men, however, could see that "the time was ripe, a rotten ripe for change, then let it come" – In fact helped it to come.[1]

[1] Excerpt from *"The Rise and Progress of the Metropolitan Fire Brigade, Melbourne, Victoria, Australia."* *"Centenary"* Brochure, by Samuel Mauger, Jr. (1934), courtesy Metropolitan Fire Brigade.

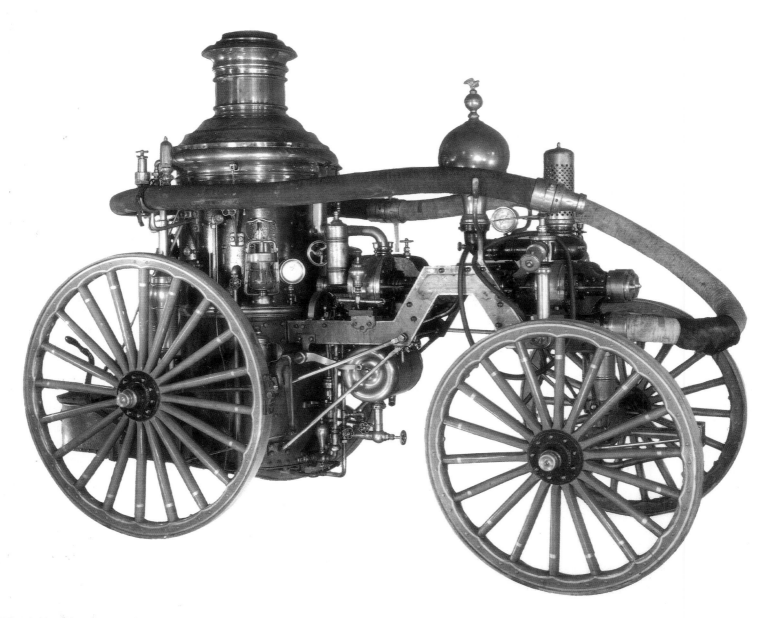

STEAM MANUFACTURERS TAKE THE STAGE

More and more builders jumped on the steamer wagon to capitalize on the burgeoning market in the United States. New manufacturers appeared throughout the nineteenth century and each tried to build a better steamer. The Silsby Manufacturing Company of Seneca Falls, New York, was famous in 1856 for its rotary pump engine and geared rotary pump engine (unlike the pistons and reciprocating pumps of other models). Silsby was the only manufacturing company to concentrate on rotary pumps and produced over 1,000 engines before joining the American Fire Engine Company in 1891. Lee & Larned, also of New York, built the "Elephant" in 1859, a rotary pump steamer with brass and steel parts commissioned by fire insurance companies for the Manhattan Engine Co. No. 8 – this was the first steamer in regular service in New York City.

Reanie & Neafie of Philadelphia built 33 steamers between 1857 and 1870 while Poole & Hunt of Maryland started building steamers in 1858, one per year for the next seven or eight years offering three sizes of chassis and enough power to produce a stream of over 250 vertical feet. These engines could also produce steam within a remarkable 11 minutes.

ABOVE: **The picture and engraving of this Silsby steamer reflect the attention to detail employed in building the average steamer, as well as an eye for style.**

RIGHT: **A Reaney & Neafie steamer, 1857, known as 'The Pioneer.' This model is from the Philadelphia Hose Company No. 1 and remains on display in Philadelphia.**

MAYER & STETFIELD'S LITH. 97 STATE ST. BOSTON.

BUI
AMOSKEAG MANUFACT

Wm. Amory Treas.
City Exchange Boston Mass

AMOSKEAG

Y THE

NG CO. MANCHESTER, N.H.

E.A.Straw Agent.
Manchester N.H.

PREVIOUS SPREAD: **This black and white lithograph of an Amoskeag steamer dates from 1861, by Julius Mayer.**

RIGHT & BELOW: **Philadelphia Hose Company Reaney & Neafie steamer, the photo taken circa 1858. According to the CIGNA Museum & Art Collection (Philadelphia), this is 'the oldest known extant American steam fire engine.'**

The Amoskeag Company of New Hampshire shifted its focus from locomotives to fire engine steamers, building its first in 1859, and developed a solid reputation. It became one of the country's top producers and put many smaller manufacturers out of business. The company's first 11 engines were considered "mongrels" because of their rotary pump with two reciprocating pistons. This ultimately proved too complicated for manufacturing purposes. The company eventually switched to single pistons, then to double-piston, double-pump models. Amoskeag produced over 1,000 engines and they were major players by the end of the 1860s (competition included Silsby in Seneca Falls, NY, Clapp & Jones of Hudson, NY and the Button Company).

Amoskeag built self-propelled steamer behemoths with the transmission chain drive on one side of the engine. This required the invention of the differential gear in order for them to make it around corners successfully, the first company to use such gears on a self-propelled vehicle in automotive history. Eventually, Amoskeag became the standard against which other manufacturers would gauge the success of their own engines. Amoskeag merged with the Manchester Locomotive Works in 1872 and prompted many smaller competitors to merge in order to stay alive, to say nothing of those manufacturers who still produced manual engines. In 1860, John Agnew of Philadelphia tried to convert his company to steam engine production but managed only four before he gave up.

The possibilities were enormous but the field thinned as the glut of manual manufacturers fell under the pressure of steam, while smaller steam manufacturers simply couldn't keep up. The export business offered a solution for some manufacturers such as Ettenger and Edmond of Richmond, Virginia, who built steamers for St. Petersburg in 1864 (up to steam in 10 minutes). Even the venerable Hunneman brand name couldn't survive faced with such competitive upstarts. Even converting to steam in the 1860s did nothing to slow the company's decent. After selling only 30 engines, the firm reluctantly closed down.

HIA HOSE CO.ˢ STEAM ENGINE

eam through 1¼ in. nozzle 210 feet — or 2 Streams, ⅞ in. nozzles, 175 feet each, Wᵗ 7,000 ℔
eam through 1½ in. nozzle 200 Feet — or 2 Streams, ¾ in. nozzles, 160 feet each, Wᵗ 6,000 ℔
Stream through 1 in. nozzle 200 Feet, Weight, 5,000 ℔

NEY NEAFIE & CO. BUILDERS PENN WORKS PHILA

—MERRICK & AGNEW'S, F

ENGINE MANUFACTORY.—

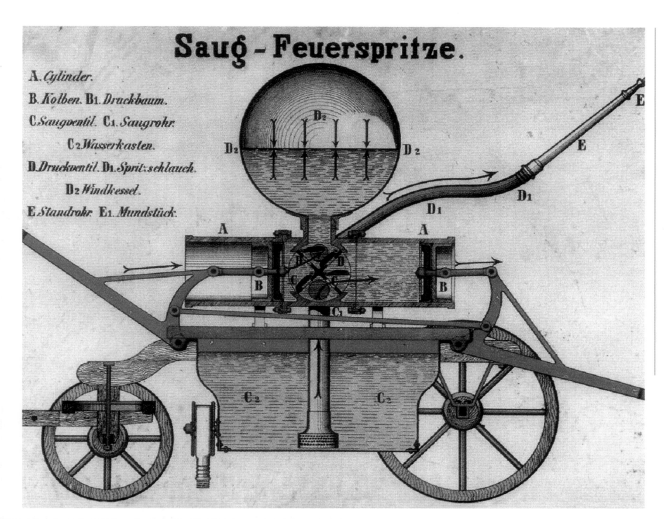

Saug - Feuerspritze.

A. *Cylinder.*

B. *Kolben.* B1. *Druckbaum.*

C. *Saugventil.* C1. *Saugrohr.*

C2. *Wasserkasten.*

D. *Druckventil.* D1. *Spritzschlauch.*

D2. *Windkessel.*

E. *Standrohr.* E1. *Mundstück.*

BUILDING A BETTER PUMP

Reciprocating pumps have formed the basis of most pumping systems since Ctesibius's air chamber. The principle is simple enough: a cylindrical shaft houses a movable plunger with valves allowing water in one end and out the other. The top of the plunger is attached to a lever or pumping handle. The lever, when pumped, pushes the plunger through the cylinder into a water basin where the difference in air pressure forces water into the plunger through a valve. Bringing the handle back drags the plunger into a plug at the top of the shaft which fits into the hollow plunger, forcing the water out a second valve. A small quantity of water is left behind, which creates a vacuum. As the handle is pumped, there is a "give and take" between water and air pressure, forcing consistent quantities of water through the system in reciprocating amounts like an average water pump.

This system was replaced by the rotary pump, a more complicated but efficient system, in the late eighteenth century. Extra plungers or pistons were added to the original pump, operated in sequence by a rotary crank. The first piston plunged into the water to bring up a supply while the next piston went down for more, followed by the next piston and so on. By the time the final piston brought water up, the first piston descended again. This produced a consistent stream of water instead of the interrupted flow of the original pumps. Steamers and many motorized vehicles used this system for years as it provided a guaranteed flow of water.

The introduction of the centrifugal pump in the middle of the nineteenth century marked the most significant change in pumping technology since ancient Alexandria. Seagrave, a chassis

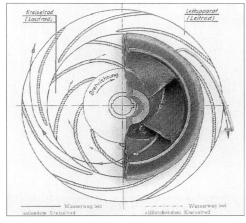

manufacturer founded by Frederic S. Seagrave in Detroit following the American Civil War, was the first to introduce the system in its earliest fire engines, around 1851. The centrifugal system brought water to the center of the pump via centrifugal force – the same sensation you felt when you're spun around by your arms. Water is moved to the outside of the pump at high speed with tremendous force, creating a vacuum which brings in more water, again forced through the system by centrifugal force. Unlike the reciprocating or rotary pump system, the centrifugal pump represented a more efficient way to pump water with far less wear. A reciprocating pump depends on gears and pistons turning against each other; it didn't take much for them to wear down over time. The centrifugal system took full advantage of physics and proved a simple and effective way to do the job.

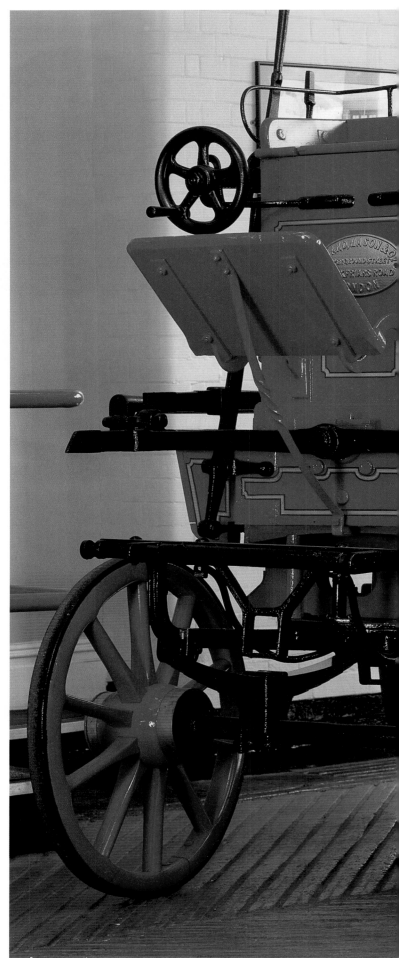

GREAT ENGLISH STEAM

The Shand Mason Company produced their first land steamer in London in 1858, drawn by three horses and weighing approximately four tons. Although the British looked unfavourably upon land steamers, they accepted steam engines in fireboats, which is where Shand Mason made their name before turning to fire engines.

The company's origins stretch back to 1774, when a mechanic named Phillips began to manufacture fire engines. The firm was taken over by a few entrepreneurial builders during its early history, including the manual pump builder Tilley, before the Shand Mason Company was officially established. The company's first steamer model was rejected in large part because of James Braidwood, the first Chief of the London Fire Brigade, who was adamantly against steamers. He argued they didn't bring the fire fighters close enough to the fire and that the water power did more damage than good. Shand Mason persevered, producing the

J.T.WOOD 278 STRAND.
THE LATE Mr JAMES BRAIDWOOD.
SUPERINTENDENT OF THE LONDON FIRE BRIGADE.
WHO LOST HIS LIFE AT THE GREAT FIRE TOOLEY ST.
JUNE 22, 1861. IN HIS 61 ST. YEAR.

ABOVE: **James Braidwood, superintendent of the London Fire Brigade, was one of the main opponents of the steamer in London, arguing them too powerful for the city.**

LEFT: **The Metropolitan Fire brigade (UK) used this Shand, Mason & Co. manual pump (1882) for years before steam came into greater use. The foldable brakes kept in compact for narrow areas and better balance.**

TOP FAR LEFT: **A combination double-piston hand pump and hose wagon by Merryweather, very useful for rural areas with a direct water source.**

BOTTOM FAR LEFT:
The proud insignia of Shand, Mason & Company.

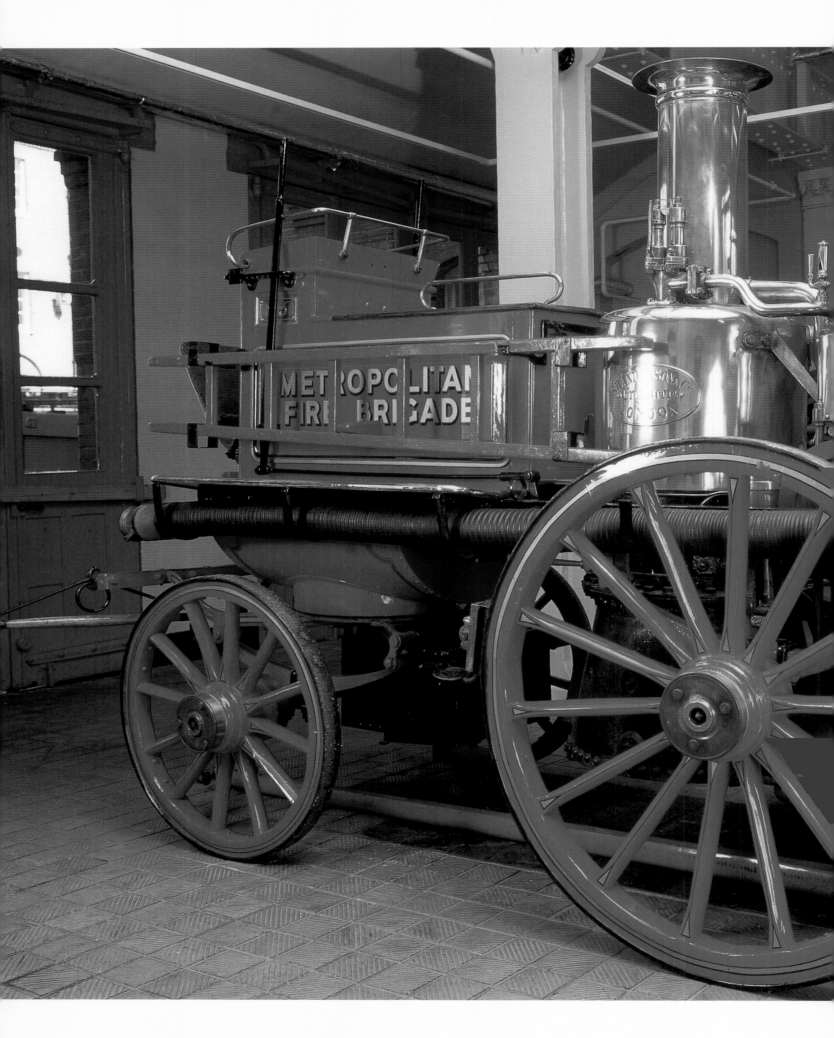

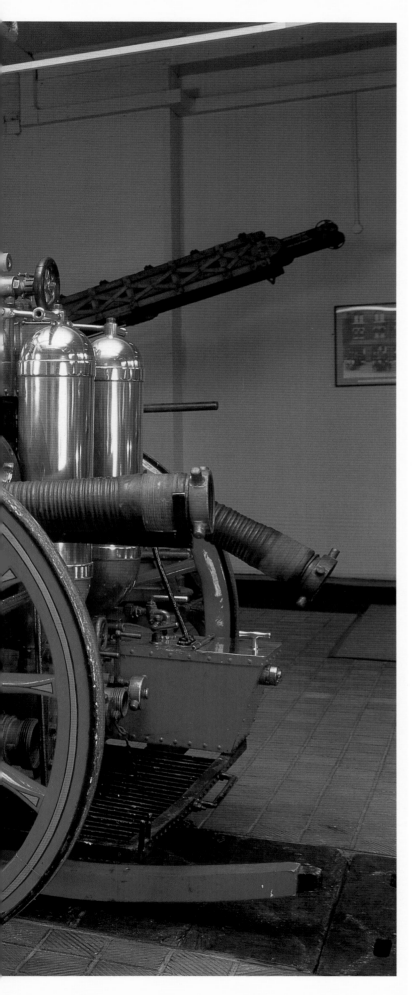

"London Bridge Vertical" in 1863, the first example of a short-stroke, high-speed engine in a fire engine (a short-stroke engine relied on short, repetitive pumps and produced a strong, more consistent stream of water). Significant advances were made to the Vertical by 1876, including a coal bunker for extra coal to maintain the steamer during longer engagements and a simple draw-pole system to release a horse if it fell en route (this was a major problem with the increased use of horse-drawn engines). More horses were required as engines grew in size, an unpredictable and dangerous arrangement. Horses often fell and were injured as they pulled these massive machines along cobblestone, uneven roads or muddy tracks. Solutions such as Shand Mason's draw-pole were vital, limiting horses' suffering and saving money for the departments.

Elsewhere in Europe and England, steamers were introduced with great success. James Skelton introduced the first steamer in Ireland in 1859 while a manufacturer named Cowan of Greenwich, England, built a horse-drawn steamer for the Library of St. Petersburg in 1863. Egestorff created a single cylinder steamer in Hanover, Germany, in 1863, Moltrecht built a similar sized engine in Hamburg, Germany, and Flaud built one for Paris, France, that same year.

William Roberts of Millwall holds the honor of creating the first self-propelled steamer produced in England. He was so proud of his engines that he entered his "Princess of Wales" steamer in the Crystal Palace trials of 1863, where engines were put in direct competition. Orders for these machines came in from as far away as Rio de Janeiro, where he shipped a middle-sized engine called the "Excelsior," very similar in design to the Princess.

James Compton Merryweather, son of Moses Merryweather, the famous manual pump manufacturer, having already built some of the most successful hand pumps in England, produced some of his best steamers through Merryweather & Sons. The Merryweather "Deluge," built in 1861, was Merryweather's first land steamer, followed by the "Torrent" in 1862 and the "Sutherland" in 1863. Merryweather built their first self-propelled steamer in 1899, but their "Fire King" models of 1902 and 1918 were among the most famous in England.

LEFT: **This Shand, Mason & Company self-propelled steamer – the 'Victoria' – was the winner of the Great Exhibition of 1863.**

ABOVE: **Merryweather & Sons steamer, 1880, built for the town of Canterbury in England.**

ABOVE: **This Waterous steamer, built in 1898 and photographed in Victoria (British Columbia), is a stunning example of the long-term quality of these engines.**

RIGHT: **The Great Engine Contest of 1850 in Philadelphia. Trials such as these were often a means for prospective buyers to see steamers in action up close.**

THE FOUNDERS

Competition in the United States weeded out lesser companies and forced those that remained to improve. Some of the most famous names in fire engine history appeared during this contentious period and in some cases disappeared, whether through mergers or financial failure. These years were among the finest in steamer history as well as the fastest. The changes came fast and furious.

The Button Fire Engine Works of New York, well known for producing some of the best manual pumps available in the United States, sold its first steamer to the city of Battle Creek, Michigan, in 1862 under the new name of Button and Blake, 30 years after its founding. It went on to produce approximately 200 steamers before 1891, at which point it joined the American Fire Engine Company, predecessor to the International Fire Engine Company, and eventually the American LaFrance Fire Engine Company. Early Button steamers had pumps mounted on the front of the chassis but later moved them to the center of the engine body for greater stability.

M.R. Clapp of Hudson, NY, a former employee of the Silsby Company, joined his associate Jones to form the Clapp & Jones Company in 1862, building fire engines characterized by a unique rubber suspension system for the rear wheels and vertical boilers. Clapp & Jones produced fire engines for 30 years – over 600 in total – before joining the American Fire Engine Company in 1891.

Lane & Bodley, meanwhile, bought the original A.B.& E. Latta factory in Cincinnati, 1863, after Alexander Latta retired. The new owners maintained the basic Latta self-propelled design, using the Latta square-boiler design with coiled tubes and three wheels. The company was sold to Chris Ahrens, the firm's former superintendent, in 1868. He focused on Amoskeag-style engines, giving up self-propelled machines and concentrating on improving warm-up time (now down to five minutes) and adding a hand-operated auxiliary pump next to the boiler so the engineer could pump manually until steam was raised. Ahrens's company produced more than 700 steam engines between 1868 and 1891, at which point it too became a part of the American Fire Engine Company.

Truckson LaFrance (formerly Hyenveux) of Seneca, New York, patented several improvements in rotary steam engines in 1871 and

1872. Seneca was already known as "The Fire Engine Capital of the World" – LaFrance certainly confirmed this reputation. The first LaFrance Manufacturing Company steamer used rotary pumps for both fire pump and steam engine, a unique idea but short-lived. The company sent a steam engine to the Paris Exhibition in 1878 but the engine failed to meet French standards for boilerplate thickness and was not allowed to participate, so the trip was useless. They paid close attention to what was happening in Europe, however. The company was renamed the LaFrance Fire Engine Company in 1880 and from that point on built engines to client specifications. This placed them among the top manufacturers in the field.

Manufacturers were not usually willing to put time or money into such specific attention to detail, preferring to produce engines en masse and as fast as possible to compete. The company produced 500 engines before joining the American Manufacturing Company and forming the International Fire Engine Company in 1900.

Waterous Engine Works Company of St, Paul, Minnesota, was founded by C.H. Waterous in Brantford, Ontario, 1884. Waterous fire engines were in such demand by 1881 that a second company was opened in Winnipeg by Waterous's twin sons, Fred and Frank. Demand soon outgrew the company's size and the operation was moved to Minnesota in 1888, where their success was based at first on small lightweight steamers perfect for rural areas. They would later be among the first developers of gasoline-powered pumping engines. They would be joined in their gasoline-fueled efforts by the Nott Fire Engine Company of Minneapolis, who built steamers in 1900 to compete with the American LaFrance "Metropolitan."

Each of these company names has come to mean much more than the machines they manufactured. They became symbols of the ingenuity and integrity that characterized this period in American steamer history, forerunners in technology. It was due to their perseverance that the steamer became one of the most vital additions to fire fighting apparatus history.

HORSEPOWER

Horse-driven fire engines came as no great surprise given the number of engines powered by horses for farm and factory use. James Watt established the relationship between the power of a steam engine and the number of horses it replaced, though there was little information on which to judge this unit of measurement. Animal-driven fire engines were much more common in rural areas where there wasn't money for steamers but there were plenty of horses available.

A standard animal-driven pump was patented by an American, John McCarthy, in 1869 (horses walked on an inclined ramp to power the pump). Another popular design was the rotary system patented by William Underwood of Louisville, Kentucky, in 1879. Horses or oxen walked on a circular treadmill to power the pump, which was effective enough but the design seemed far too large and awkward for its purposes. Benjamin Howe produced a similar rotating horse engine in 1881 with a large horizontal cog wheel and a central lever to which a horse was attached. These engines did not achieve the status their innovators hoped, but they proved interesting nonetheless.

TRIAL AND ERROR

Steamer manufacturers exploited the competitive spirit born with the manual pumpers to the fullest. Increased competition prompted trials and competitions, a chance for steamers to test their strength and prove their dominance over manual pumps.

New York City was the scene of many competitions, perhaps providing the last nail in the coffin for the New York Volunteer Fire Department. As Costello describes it,

> The old hand apparatus, made as light as possible to permit of quick hand transportation and elegance of construction, had to hug a fire to permit a proper stream being thrown on the flames by ordinary exertion, which, as a rule, was no light work, under extraordinary exertion, and the stimulus of rivalry with the iron-lunged innovation, streams were thrown which equalled, if they did not excel, those thrown by steam power. But it was at once seen that as a rule the steam machine could take a remote hydrant and do as good service as a hand machine, until the men who manned the brakes became exhausted, and then continue to do equally good service as long as fuel lasted. Besides, it was evident that the resources of science and mechanism had not been exhausted in the construction of the first steam engines.

The city council voted to do away with the volunteer department and institute a paid department, to the dismay of the volunteers. Trials provoked tremendous rivalry but they were also a solid gauge of the relative success or failure of any particular engine.

RIGHT: **The Great Fire of Chicago (October 8, 1871). As with the Great Fire of London, boats gather at the waterfront as people try to escape the flames.**

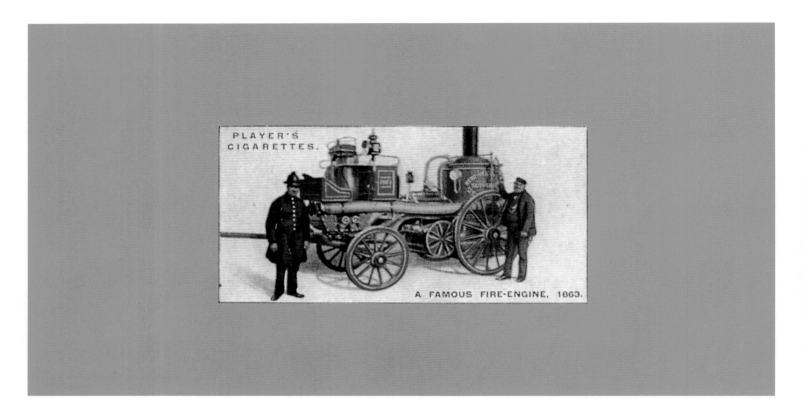

A FAMOUS FIRE-ENGINE, 1863.

PLAYER'S CIGARETTES.

The Philadelphia Steamer Trials in 1859 were one of the more colourful tests devised and reinvented the typical "how high, how far, and how strong can you go?" tests of the day. A mast 173 foot high was erected in the main arena with tin cups arranged at varying heights. Competitors had to fill the cups with their engines. The winners were the "Hibernia" by Reaney & Neafie of Philadelphia and the "Washington" by Poole & Hunt, and competition was fierce.

Tests were not limited to the United States and competitors were invited from all over the world. The International Exhibition in London (1861) for example, was supposed to pit the Merryweather Deluge against a Lee & Larnard five- horsepower manual hand pump from New York, but Lee disagreed with the standards set by the judges and refused to take part. Two Shand Mason steamers were sent instead, more for show than any other reason, and both fared very close in their results.

At the Crystal Palace in London two years later one of the more famous trials took place. Ten entrants (seven British and three American) competed. Lee & Larnard's "Manhattan" was there, undeterred by their first questionable London experience. This optimism was crushed by an accident the Lee & Larnard engine suffered at the hands of the London Fire Brigade – engines were tested by random teams to keep things fair but the London Fire Brigade "accidentally" crashed the Lee & Larnard engine on the way to the testing ground. The accident was later blamed for Lee & Larnard's failure at the trials. Merryweather had two steamers in the competition, the Torrent and the Sutherland, the latter taking first prize for the large-engine class and the former receiving honours in the small-engine class. The integrity of the English steam engine manufacturer was secure despite the country's late start in steam engine production. The Lee & Larnard team would probably cry foul at this assertion, voicing the same accusation as they did during the previous trial, that the judges were not experts. Trials were often the easiest way for towns and cities to test a manufacturer's work, as was the case with the Rotterdam Trials in Germany and the

ABOVE: **A cigarette card illustration of the Merryweather & Sons 'Sutherland' that won the first prize at the International contest at Crystal Palace in England.**

RIGHT: **Plaque from the 'Victoria' vertical steamer by Shand, Mason & Co. – including prizes and prize money won.**

Middleburg Trials in England, 1864, in which a single cylinder, vertical pump Shand Mason steamer competed with a double-cylinder, horizontal pump Merryweather. Money was invested to bring engines over and set up the competition but the spectacle was worth the effort and there was no better way to check out a particular engine than seeing it on your own home turf.

MERGERS

Necessity breeds change. Smaller companies joined forces with other small companies to increase production, add an element of fire engine production that was lacking, or compete with larger companies. In what may have been the catalyst for the mergers to come, Amoskeag of New Hampshire was incorporated into the Manchester Locomotive Company in 1872. The "Fire Engine Division (Amoskeag) of the Manchester Locomotive Works" had proved its worth as a manufacturer of some of the best steamers available. As part of a larger, more financially powerful union, it represented a real threat to any company hoping to compete.

Chris Ahrens spearheaded an amalgamation of his own to keep apace. The Ahrens Manufacturing Company, the Button Fire Engine Works, Silsby Manufacturing Company, and Clapp & Jones formed the American Fire Engine Company of Cincinnati and Seneca Falls, in 1891. The company established itself as an industry leader, building the famous "Metropolitan" steamer with its flywheel placed between two vertical cylinders, and it seemed it would stay at the front of the pack for some time. The company's decentralized structure, however, kept each sub-company independent and, while

this sounded good on paper, it prompted the amalgamated company's eventual dissolution.

Ahrens did not give up, however. He kept the company's name and teamed up with the Fire Engine Division (Amoskeag) of the Manchester Locomotive Works, the American Fire Engine Company, and the LaFrance Steam Engine Company, plus six smaller companies (Charles T. Holloway & Co., Fire Extinguisher Manufacturing Co., Rumsey and Co. Ltd., Gleason and Bailey Engine Works, Thomas Manning, Jr. and Co., Macomber Chemical Fire Extinguisher Co.) and formed the International Fire Engine Company in 1901. It was renamed the American LaFrance Fire Engine Company in 1904 to further consolidate the companies. It didn't last long. Ahrens, apparently unhappy with the company, broke off to form the Ahrens Fire Engine Company with the help of his sons Fred and John as well as his sons-in-law Charles Fox and George Krapp, producing a new model called the "Continental." In 1908, Charles Fox became president and the Ahrens Fire Engine Company, the now famous Ahrens-Fox.

Amalgamations gave smaller companies the financial backing to grow in the United States but also limited the field to only the largest or the most specialized companies. With the development of motorized fire engines, however, even the largest of companies could fall, if they were not prepared to face the reality of a fast-changing world.

STANDARDIZED SIZES

The road leading to the development of motorized engines was not a straightforward one. Discrepancies in claims for engine power (gallons per minute) prompted the National Fire Protection Association's chairman, Captain Greeley S. Curtis, to establish a standardized set of measurements for steamers that allowed for more accurate comparative assessment, in 1906. Measurements were standardized for all manufacturing companies, based on three sizes (600, 800 and 1,000 gallons) and not the eight to ten sizes that were used by most manufacturers. These three sizes represented common ground for manufacturers. Unfortunately, engines were often produced in very different sizes with various pumping systems of varying strengths, making standardized measurements difficult to enforce.

These standards were an important step before the introduction gasoline-powered engines. Accurate standards were the perfect ground in which to plant the seeds for the gas-powered engine, separating lesser engines from the better. As the dawn of the twentieth century broke, the picture of the future was becoming clearer and clearer. These steps were an essential part of that future.

LEFT: **The American LaFrance 1918 motorized fire engine – a shiny sign of things to come.**

4

Manufacturers Start Their Engines

FIRE FIGHTING CHANGED PERMANENTLY WITH

THE DEVELOPMENT OF THE GASOLINE-POWERED

INTERNAL-COMBUSTION ENGINE. STEAMERS HAD

IMPROVED QUICKLY DURING THE COURSE OF

THEIR SHORT HISTORY BUT NO-ONE COULD HAVE

PREDICTED THE FAR-REACHING EFFECTS OF

MOTORIZATION. HORSE-DRAWN STEAMERS

BLAZING DOWN CITY STREETS, ADMIRED AND

CHASED AFTER BY EXCITED CROWDS, WERE SOON

REPLACED BY THESE STREAMLINED, ECONOMICALLY

VIABLE MOTORIZED VEHICLES. TIME AND MONEY

WERE OF THE ESSENCE AND THESE NEW MACHINES

MET THE CHALLENGE HEAD ON.

The Brave New Motorized World

Most North Americans associate the internal combustion engine with Henry Ford and the mass production of automobiles. The earliest developments in combustible engines stemmed from Europe however and not the United States. The U.S. only took to the fore once again with the eventual streamlining of these engines.

The story of the internal-combustion engine begins in France and Germany, spreading across the nineteenth century in spurts of inspiration. Sadi Carnot established the theory behind an internal-combustion engine in France, followed by Alphonse Beau de Rochas, who built on Carnot's work to come up with the four-stroke engine. This standard piston system formed the basis of the internal-combustion engine. A German by the name of Nikolaus August Otto earned fame for this idea around 1876, but couldn't claim the idea

of a four-stroke engine as his own (although he did try). The four stages of the "Otto Cycle" used the gases created by an explosion of gasoline and oxygen to power a piston. The same piston once used in steamers was now powered by very different and more effective gaseous mixture. The four stages were intake, compression, power, and discharge. The cycle would then repeat and the piston would get the engine moving. By exploiting the true power inherent in fire, these early inventors were able to create something even more powerful than steam to take on the fire, although it would be a few years before anyone realized how this new invention might be applied to fire fighting.

The origins of gasoline-powered fire engines is a bit confusing. Some pumps were themselves pumped by an internal-combustion

LEFT: **A 1929 Ahrens-Fox, offering 1000 gallons per minute (gpm).**

PREVIOUS SPREAD: **A 1924 Whippet chemical engine, fully-restored.**

engine while the vehicle was horse-drawn. In other cases a steam engine pump was mounted on a gas-powered engine chassis for easier transportation. Each approach was significant, but it is difficult to give credit to the true pioneers. It also becomes more complicated to assign predominance to one side of the Atlantic over the other as both had a part to play. Otto for example received a patent for his engine design in the United States and built stationary engines beginning in 1878. In 1885 in Germany, Carl Benz built the first gas-powered internal-combustion engine, set in a three-wheeled vehicle. Two former employees of Otto's, Gottlieb Daimler and Wilhelm Maybach, concentrated on applying the technology to something a little more mobile. They had mounted the first high-powered internal-combustion engine onto a four wheel chassis by 1886. By 1889, Emile Levassor, a licensee of Daimler's technology, had built the first automobile, creating many of the staples of car manufacturing that we still enjoy today, like steerable front wheels and a front-mounted engine.

Not everyone liked the idea of a gas-powered pump and engine. Daimler, for example, eventually shifted his attention elsewhere, concentrating on an electric engine, producing an electric-gasoline pumper with Porsche in 1914 with a pump mounted in the center of the appliance and a detachable hose reel. Justus Christian Braun of Nuremburg, Germany, created the "elektromobil-steam" fire engine, an electric vehicle, in 1910. It wasn't popular for long in the face of successful motorized engines. This machine was unique if only because, in addition to the electric battery powering the vehicle (and acting as a driver's seat), the boiler was supplemented by a gaseous combination of kerosene fuel and carbonic acid.

From the changes in motorization to experiments in electric automobiles, the turn of the century marked a stage of ingenious though somewhat disparate development in all areas, as designers, engineers, and inventors competed with one another to find the Golden Fleece of motorized vehicles. As with steam, the public would be the final judge.

THE EARLY YEARS

It wasn't long before the first internal-combustion engine was fitted to a fire engine. There are claims to the honor on both sides of the Atlantic. The first rests with the Waterous Engine Works Co., Ltd. of Minnesota, responsible, in 1898 for the first gasoline-powered pumping engine built in the US. This engine was still not self-propelled, however. It was mounted on a horse- or hand-drawn chassis and was similar in model and intent to the steam engines that preceded it, the only difference being the source of the power. Waterous produced a gas-powered self-propelled fire engine and pump in 1906 but it still needed the help of horses or brute strength to get to an engagement.

Merryweather & Sons are credited with the first self-propelled gas-powered fire engine put into service by a public fire department in England, commissioned in 1904 by the Finchley Fire Brigade. The engine and its pump were both gasoline powered, unlike earlier experiments with gasoline-powered engines where either the pump or the vehicle itself were gas-propelled but not both. It had a 30-horsepower engine capable of 575 gpm and a top speed of between 10 and 15 kilometers (6¼ to 9¼ miles) per hour.

Waterous and Merryweather are pinpointed as the "first" producers of these engines but they achieved this status more by chance than due to specific innovation. Merryweather for example produced its popular Fire King steamers until 1918 and they sold well throughout those years. It did not commit to gasoline-powered engines until it was obvious there was no turning back. The technology for gasoline-powered motorized engines was available to everyone by the turn of the century. It wasn't long before a slew of manufacturers, both new and old, took the global stage with their own unique contributions.

Delahaye-Farcot of France produced a large motorized engine in 1907, which was very popular both in France and abroad. It carried hose reels, pumping equipment, and up to 15 fire fighters. Their work was so effective, Delahaye engines made up the majority of the appliances used by the Corps de Sapeurs Pompiers by the end of the 1920s. Following the same pattern, Isotta-Fraschini of Milan, Italy, built an engine for the Turin Fire Brigade in the early 1900s

ABOVE RIGHT: **This model combines a 1930 Ford AA-Barton engine with a 1912 Seagrave body (250/100).**

RIGHT: **A motorized combination first-aid appliance from 1903. According to the cigarette card on which this Merryweather (UK) model is illustrated, it was one of the "first petrol motor fire-engines."**

PLAYER'S CIGARETTES.

"FIRST-AID" APPLIANCE, 1903.

with a rear-mounted pump and crew seats facing outward. The Braidwood body style of the previous century had inspired many aspiring manufacturers. Braidwood, the captain of the London Fire Brigade, was the first to sit firemen on top of engines, with room for four to six men. Fortunately for the crews of both engines, these early models were relatively slow. Engines, however, became more powerful and grew faster. Rough roads and sharp turns were enough to send more than one fireman flying.

Vaclav Laurin and Vaclav Klement of Jungbunzlau, Czechoslovakia, were successful builders of early fireman-carrying fire engines. They began their careers as motorcycle and automobile manufacturers before turning to fire appliances in 1913. They offered a six-liter engine with 32 horsepower capable of pumping almost 40 gpm. These were complete machines, carrying hose and ladder as well as six members of the fire fighting crew. The two Vaclavs built their seats to accommodate the turns and jostling of the engine. This engine design was eventually replaced by enclosed cabs but while it was used it made traveling easier and safer for everyone – although the fire fighters still had to hang on for dear life.

FIRST STEPS FOR A NEW ENGINE

As former locomotive engineers and companies turned their eyes to the potential of the steam fire engine, car manufacturers tried their hands. The most famous name and the father of mass production worldwide was the Ford Motor Company, established around 1908. The Ford Motor Company established itself as the leader in the mass production of motorized vehicles with its very popular and remarkably inexpensive Model T. Many smaller communities used a Model T coupled with a small gasoline-powered pump to produce an inexpensive fire engine. This allowed communities with few resources to put together their own fire engines at a relatively low price. Ford introduced the Ford Model TT in 1917, which gained popularity as an inexpensive proposition for putting together a fire engine. By 1927, buyers were anxious to see what Ford would do next and anticipated the Model A engine eagerly. The National Board of Fire Underwriters tested Ford's new model and it was everything the public had hoped for and more. This style of engine was used regularly across the United States incorporating some of the best pumping engines available.

LEFT: A 1938 Ford/
Darley/Memco
(65 foot ladder),
with a distinguished
enclosed cab.

BELOW LEFT:
Julius Pearce provided
the body for this
1922 Ford-TT hose
and chemical
motorized engine.

engineer while another used a Rolls-Royce chassis converted to fire fighting with a familiar front-mounted pump – this model was sent to Hong Kong during the years before World War II.

Ingenious thinking like this led to the first practical safety measures. Dennis built one of the earliest incarnations of the enclosed cab in the 1930s (the Dennis FS6, for example, seated 12 men quite comfortably). Their enclosed cab was based on a 1929 design by a Glasgow brigade driver called the "New World Body" or an "all-weather pump." It included a very low center of gravity and chassis for additional safety. Innovation kept Dennis at the forefront of the industry and contemporary Dennis engines still fly through the streets of Britain on the way to a fire, still proving their worth after almost a century of manufacturing.

Manufacturing took off in the United States with the replacement of the horse-drawn chassis. This was not easy. Fire fighters liked their horses and the public was used to horse-drawn fire engines barreling down the road toward a blaze. The idea of replacing the ever-dependable horse with these newfangled horseless carriages was ridiculed in some quarters but economics were more important than an emotional attachment to a stable of horses.

The transition from steamers to motorized fire engines was quick compared with the trials and fights that raged before the steamer could replace manual pumps. Progress was less frightening and people were willing to accept new ideas in engine development. This acceptance may have been simple economics. Paid fire departments were everywhere but they were costly and needed to be streamlined. Mass-produced gasoline-powered engines were a more economical option, requiring less manpower with more effective results – the day of the horse-drawn steamers was coming to a close. The birth of the mass-production manufacturer was at hand.

This was true in England as well as the United States. The Dennis Brothers built up quickly and were successful, just by meeting the demand. Their first appliance was delivered in 1908, sold for £900, and included a single-stage centrifugal pump bought in London and a front-mounted pump. They were one of the first to include developments in their designs from around the world, as well as one of the most prolific innovators in the field, making their final product much stronger. They were not afraid to try something new in order to meet the demand at an affordable cost. One later Dennis model, for example, included a turbine pump designed by an Italian

THE AGE OF THE AMERICAN MANUFACTURER

Despite public resistance to change, the rise of motorized vehicles was the dawn of a new day for American fire engine manufacturers. Ford led the world in mass production and it wasn't long before other American manufacturers followed, setting up operations throughout the United States that far surpassed those of any other country both in production and quality. Most fire engine manufacturers that came of age during these years enjoyed tremendous success and volume business, in some cases on an international scale. This was the golden age of the American fire engine.

CHRISTIE'S FRONT DRIVE AUTO COMPANY

John Christie's company, called Christie's Front Drive Auto Company (New Jersey), though originally a producer of tractors, realized the opportunity of fire engine manufacturing early. Christie created a two-wheeled, four-cylinder tractor to replace the horse-drawn steamer in 1911. There was no urgent need to replace steamers with gas-powered pumps but horses were expensive. The choice paid off and Christie's produced over 600 tractors during the next eight years. The company closed down shortly after World War I and Christie eventually achieved fame for armored tank design.

MACK BROTHERS COMPANY

Despite public resistance to change, the rise of motorized vehicles was the dawn of a new day for American fire engine manufacturers. Ford led the world in mass production and it wasn't long before other American manufacturers followed, setting up operations throughout the United States that far surpassed those of any other country both in production and quality. Most fire engine manufacturers that came of age during these years enjoyed tremendous success and volume business, in some cases on an international scale. This was the golden age of the American fire engine.

The company outgrew its home turf quickly and moved to Pennsylvania in 1904 to larger headquarters. Mack entered the fire engine arena only after perfecting its own engine – it added a mounted Gould steam pump to its own chassis. It wasn't a real Mack fire engine, just a quick gas-powered replacement for horse-drawn vehicles. The pump was placed on one of the Mack Senior Series heavy-duty truck chassis and sold to the Union Fire Association in Philadelphia, where it was received with very high acclaim. The truck was of the highest quality and established Mack as one of the best heavy-duty vehicles for fire engine use. The small Mack AB model chassis, built in 1914, was the basis for the later AC truck, which sold more than 50,000 machines between 1914 and 1937. Mack's Model AC "Bulldog" truck was first introduced in 1915 (World War I doughboys named the truck for its tough, canine appearance). Its distinctive pug nose became the model for the majority of Mack trucks in years to come. The AC was most popular during World War I and more than 40,000 AC Bulldogs were manufactured through to 1938. The durability and reliability of these machines was never questioned. Surplus Mack trucks returning to the US after the war, trucks that had been put through rigorous paces and had faced battle after battle already, were converted to fire engines.

K-1543

ABOVE: **A pre-war ladder truck. Mack was well-respected largely due to its distinguished performance under war-time conditions.**

RIGHT: **The bulldog logo – derived from the trucks' nickname during WWI – gave Mack a memorable brand name and an extra edge over the competition.**

Many Mack trucks never made it back from the war. Europeans recognized the strength of these tremendous engines and kept them for their own cities when possible. Mack recognized a need that was not being supplied and added the Mack 500 gpm pumper to its roster in 1915. Mack wouldn't limit production to fire trucks and pumpers of course. The company introduced an engine-driven mechanical hoist for its aerial ladders by 1928. It created a stir in 1935 with the first American enclosed sedan-type pumper, built to the specifications of a fire department in North Carolina. Their enthusiasm was obvious. The cab enclosed almost everything, only small lengths of hose visible outside. Equipment, firemen, and most of the hose were contained within the body of the vehicle, amusing both spectators and fire fighters and sparking the creation of the 1936 Mack E Series in line with expectations. This model made a lasting impression.

The Mack truck remains the archetypal image of a fire engine in North American memory, a strong image even with the advent of flat-fronted forward cab design. The 1938 Type 80 truck introduced this "look." It was powered by a Mack 168-horsepower engine and carried the distinctive front grille and bulldog mascot. This was the image of a Mack truck and it remained its trademark through World War II and into the 1950s.

LEFT: **A 1938 Mack engine offered only 150 gpm but was still a sturdy and reliable machine.**

BELOW LEFT: **A 1935 Mack model AP-B (750 gpm), which includes a restored water tank – the original was much smaller.**

BELOW RIGHT: **A 1927 Mack, with the open cab and Mack's distinctive front grille.**

1: American LaFrance
Type 12 from 1920,
supplied 1,000 gpm.

2: A unique
combination of a 1916
Seagrave tractor with a
1907 American LaFrance
second size steamer.

AMERICAN LAFRANCE

Mack and Christie were new faces in fire engine manufacturing but they were not alone. Some very familiar older companies made a successful transformation to gas-powered engine production – though some were more successful than others. American LaFrance was an amalgamation of numerous steamer manufacturers established at the end of the previous century. Converting to gasoline-powered vehicles was not easy. They made an attempt in 1907 but failed to impress the intended buyers and the engine was returned three months after delivery – it may have been a matter of economics though as the Boston Fire Department hadn't paid for the engine. The company was more successful in 1909, delivering two motorized hose and chemical wagons but these engines were not set on chassis built by American LaFrance.

The company's 1910 combination chemical and hose car was its official entry in the fire engine field but it didn't hit its stride until it began using the rotary-gear system. The American LaFrance rotary gear engines injected some badly needed enthusiasm in the company and marked a shift in its direction. Its 1916 centrifugal

3: **Stylish motorized engines, such as this 1915 American LaFrance, remain popular showpieces.**

4: **A 1916 American LaFrance, providing 500 gpm.**

5: **An American LaFrance model dating from 1922.**

6: **Another 1922 American LaFrance, capable of 500 gpm.**

7: **The early American LaFrance models, such as this one from 1927, were popular both for their quality and their styling.**

8: **A 1931 American LaFrance quad.**

9: **American LaFrance 1931. After two decades of producing motorized vehicles, ALF, as the company was also known, was still at the head of the pack.**

pump and later piston pumps would have the same effect as American LaFrance, or ALF as it became known, tried to keep up with changing times. The company's last steamer was delivered in 1911, a four-cylinder tractor with a 75-horsepower engine.

The 1915 Type 40 pumper and the 1916 Type 14 (later offered as a complete unit with full ladders, hose, a volume pump and a booster pump or chemical equipment) reflected ALF's commitment to quality and development. The 1922 Type 75 offered six cylinders, 105 horsepower and a 750 gpm pump. The Series 100 "Metropolitan" debuted to critical acclaim in 1926, followed by the the Series 200 "Master" in 1929, the Series 300 in 1933 (noted for its outstanding style), the Series 400 in 1935, and the Series 500 in 1938. American LaFrance accepted only the best. The Series 400 Metropolitan in particular was outstanding, with massive pumping capacity and great style. American LaFrance added V12 engines (multi-cylinder, capable of great power and speed, introduced in 1931) to its muffler-less Metropolitans. The result was loud, audacious, impressive, and very noticeable. These streamlined, powerful engines set the standard by which most fire engines in North America were judged for years.

SEAGRAVE

Seagrave stepped onto the stage in the middle of this streamlined, style-conscious battle for dominance and walked to the front of the fire engine fashion parade. Seagrave trucks recalled the days of fire fighting pride, parades down main street and memories of the flashiest engines. The company was founded by Frederic S. Seagrave in Detroit in the years following the American Civil War and eventually settled in Columbus, Ohio. Seagrave bought 36-horsepower engines from auto manufacturers Frayer-Miller set on reinforced chassis, and entered the fire engine fray. Seagrave introduced its first tractor-trailer aerial ladder rig in 1908 and established itself as a trendsetter in fire engine fashion. It began selling its Model AC in 1910, a combination chemical and hose wagon featuring the company's "buckboard" style (so named because it looked like a buckboard wagon). It produced its first pumper in 1911, enclosing a centrifugal pump in the rear under a sloping hood, the first of Seagrave's style-conscious design touches.

Style was not all that mattered to Seagrave. It introduced an aerial ladder truck with a straight frame in 1911, a hose wagon (the Model AC 80) and a truck that incorporated an automotive-style hood. Innovation was taken even further in 1912 with Seagrave's more powerful centrifugal pumps (they ranged from 600 to 1000 gpm, more powerful than ever) as well as a pressure regulator, followed by a unique self-contained auxiliary cooling system in 1915.

Seagrave's attention to style and detail was its most memorable feature. The introduction of a gabled hood to its trucks in 1913 gave Seagrave engines a look all their own. This look changed in 1921 with the addition of a rounded hood and radiator grille and this new look became a staple in fire engine looks. Mack may have been famous but Seagrave was the fashionable fire engine.

The company went international in 1929, opening a Canadian factory. It introduced its first full line of fire engines between 1923 and 1926: the "Suburbanite" (smaller and very popular), the "Special" or "Standard" (middling size) and the "Metropolite" (large). Seagrave was well on its way to becoming a household name. It offered trucks with V12 engines by 1935 to compete with America LaFrance's model and produced functional enclosed cabs by 1936. The Canopy Cab, or Safety Sedan as they were known, was another step in Seagrave engines' stylish history.

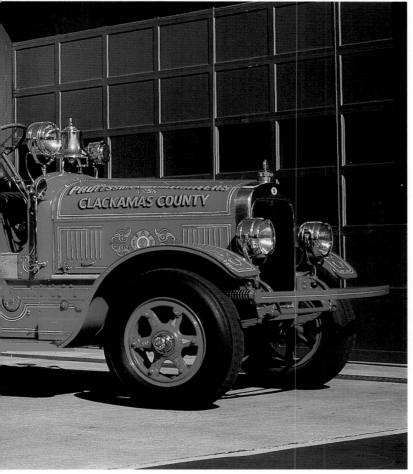

1: **A 1927 Seagrave model 6BT, producing 350 gpm.**

2: **A 1924 Seagrave, soon after the addition of the rounded hood and distinctive grille which marked Seagrave's style.**

4

4: 1935 model Seagrave.

5: 1925 Seagrave with 65 foot water tower.

6: 1931 Seagrave, offering 500 gpm.

7: Seagrave model 6DT from 1934.

8: The 1935 Seagrave, producing 600 gpm.

9: 1937 Seagrave model 80.

10: This 1938 Seagrave carried 65 feet of ladder.

5

6

AHRENS-FOX

Ahrens-Fox was at a crossroads in 1912. The company was still centered in Cincinnati and manufactured whatever the market demanded, from hand pumpers to gasoline-powered pumps. As other companies switched to gas-powered vehicles, the inevitable had to be asked – where do we go from here? Ahrens-Fox took slow and cautious steps toward all-out motorization, beginning with the 1913 "booster tank." This gas-powered centrifugal pump was set in front of the radiator on a small commercial steamer with a small water tank just behind the driver. It began pumping water before the engine was up to steam and later replaced the chemical extinguisher in most standard Ahrens-Fox pumpers.

The company's next step was its biggest leap in style, the famous front-mounted pump which appeared on the company's 1914 Model K pumper. This added esthetic element gave the engine a distinctive

look, but did not appear as streamlined as the competition. The bulk and complexity of the average Ahrens-Fox model might have unnerved potential buyers but what it lacked in subtlety it more than made up for in quality. Ahrens-Fox remains among the most respected names in fire engine manufacturing history.

Ahrens-Fox's 1919 Model J was the first modern triple-combination pumper, incorporating a booster pump for a continuous flow of water, a main pump for high-volume discharge, and a hose for easy distribution, all in one package. The Model J also had great power and style with 750 gpm capacity, a beautiful bronze piston pump, and a striking silver dome fronting the vehicle – it was considered the Cadillac of fire engines until its close in 1957.

Ahrens-Fox's best years fell between World Wars I and II when it produced some of the most impressive equipment available. Quality did not come cheap, but it was worth the expense. The Seagrave

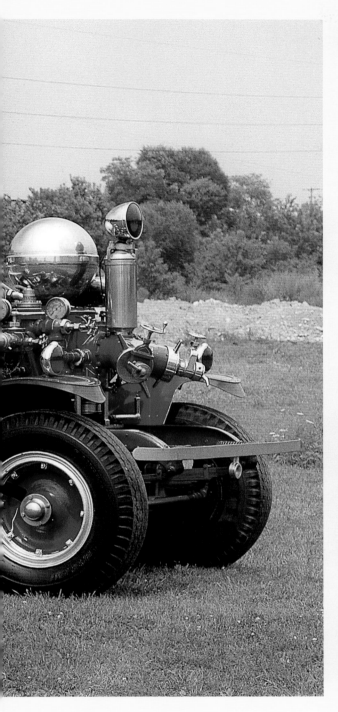

1: Ahrens-Fox was always recognizable for its front-mounted pump, as in this 1924 model.

2: 1928 Ahrens-Fox, capable of producing 1,000 gpm.

3: 1931 Ahrens-Fox quad.

4: 1929 Ahrens-Fox quad – note the reels of hose on top of the engine.

5: A shiny 1939 Ahrens-Fox, including a rounded enclosed cab top.

6: Ahrens-Fox produced this 750 gpm model quad in 1937.

pump's air chamber was transformed from one- to two-piece construction in 1923. The company introduced its first city-service ladder trucks in 1927, and stacked the ladders in two piles for safer travel in crowded streets and at higher speeds – the two piles leveled and lowered the engine's center of gravity as it rushed to a fire. Ahrens-Fox produced the "Skirmisher" around 1929, a small quadruple rig, to compete in the lower-priced market against other successful models like Seagrave's Suburbanite. The attempt was not altogether successful. Ahrens-Fox was still innovative but expensive. Costly experiments in rig design combined with the Depression forced the company into liquidation in 1937 though reorganization kept them in business until 1957.

Always faced by an uphill battle, Ahrens-Fox was never able to achieve the same status, however, and was eventually forced to close. This was not an uncommon fate, but tinged with particular sadness for Ahrens-Fox, a company that had brought such distinction to the field.

7: **A 1938 Ahrens-Fox (750 gpm).**

8: **A 1933 Ahrens-Fox C-T 4, producing 250 gpm.**

9: **A 1935 Ahrens-Fox.**

ABOVE : Pirsch, though not a major player, produced notable engines, such as this enclosed cab 1946 model.

1: 1930 Maxim B50 (500 gpm). Maxim struggled but did achieve success, particularly with its compact Magirus ladder.

2: A small but well-known company, Buffalo remains popular in parades and exhibitions around the united States. Pictured here is a 1934 Buffalo (500 gpm).

PETER PIRSCH & SONS

Peter Pirsch & Sons was founded in Wisconsin in 1857, as the Nicholas Pirsch Wagon and Carriage Plant, manufacturers of ladders and chemical apparatus. The company produced its first triple combination pumper in 1916 but did not use its own chassis for some years. This model was a triple combination machine built on a White Motor Company truck chassis and was fairly successful. Not only did the company use the commercial chassis of other companies, it never designed its own engine and stayed with the Waukesha six-cylinder fire service model instead. This made it easier for the company to adapt its work to client demands faster and with greater efficiency. This continued until the early 1920s. Pirsch began to produce its own chassis in 1926, distinguished by their white, hard suction hoses.

Like all truly memorable engines of its day, Pirsch tried to stand out from the crowd. It offered the first American apparatus with a fully enclosed cab in 1929, the safe and stylish design contrasting sharply with the rugged reputation of the American fireman in the United States. It delivered these closed cab machines fairly regularly starting in the late 1930s, often with commercial chassis.

Pirsch was no stranger to innovation either. When a tunnel collapsed in Chicago in 1931, the word went out that there were people trapped inside. Pirsch had literally just perfected a pump that could draw smoke out of buildings and tunnels, based on an idea by Minneapolis fire chief Charles W. Ringer. The engine was brought to the site of the emergency, its paint still wet, and saved the lives of 16 men trapped inside by clearing the tunnel of smoke. The device was later known as the Pirsch-Ringer smoke ejector and could be used to blow out grass fires and remove poisonous fumes as well. Pirsch's reputation for creativity and ingenuity grew not only for its work on engines but also for its aerial ladders, hose shutoff valve, and door openers.

MAXIM, NOTT & OBENCHAIN-BOYER

The Maxim Motor Company of Massachusetts, W.S. Nott of Minneapolis, and Obenchain-Boyer of Indiana were some of the many small but effective fire apparatus manufacturers spread across the United States. The Maxim Motor Company, founded by Carlton Maxim, produced its first fire engine (a hose wagon) in 1914. It was built on a Thomas Flier chassis although Maxim would build its own chassis after 1916. The company prospered during World War I in particular and offered its own line of fire engines by 1918. Its 1921 M series of pumpers and 1927 B Model were both successful engines but much of the company's notoriety was based on its "Magirus," a compact aerial ladder that found its most enthusiastic buyers in the crowded streets of many European cities (even Mack rigged its trucks with the line at the time).

W.S. Nott was founded in 1879 as a steam pump manufacturer. The company produced its first motorized equipment in 1912, reaching a peak with its "Universal" pumpers which were more streamlined than their competition and used worm drive transmission. These engines were particularly popular with the New York Fire Department in the years leading up to World War I.

The story of Obenchain-Boyer was born, appropriately enough, out of fire. The company, originally a mill operation, moved from one mill site to another in 1897. The mill was destroyed by fire a short while later, prompting Mr. Obenchain to develop and patent a

3: **The Front Wheel Drive company, also known as FWD, kept up with the pack with models such as this 1961 model.**

4: **A distinctive Ward LaFrance from 1972, with a open cab, generated 1,500 gpm.**

5: **1934 Buffalo, producing 500 gpm.**

6: **The 1962 Ward LaFrance, again with the open cab and front wind-shield.**

7: **Four Wheel Drive from 1952.**

new fire extinguisher. This led to hand-powered chemical engines in 1905 and the creation of the company's first motor-driven appliance in 1916, mounted on a Ford Model T chassis. The company reached a peak during both World Wars, like most companies. Obenchain-Boyer was considered a leader in hand-drawn chemical apparatus and supplied many motorized vehicles to the army regularly.

The history of motorized fire engines in the United States has always been varied but consistent. Developments passed from one coast to the other with greater and greater speed just as innovations from overseas found their way to the American shores in the early decades of the twentieth century. This was the golden age of motorized fire engines both in America and in Europe. Four Wheel Drive, Stutz, the Buffalo Fire Extinguishing Manufacturing Company, Teudloff Dittrich, Howe, Ward LaFrance, Laffly, Sanford Fire Apparatus Company, Moreland Truck Company, Scemia, Brockway Truck Company, Leyland, Raba Fire Engines, and Rosenbauer Appliances were only a few of the well-known names on the market and the list was still growing. More than a thousand years had passed between Ctesibius's air chamber and it's re-introduction. It took less than half a century for the motorized fire engine to take over.

LADDERS STEP UP

With the advent of both electric and gasoline-powered vehicles, aerial and turntable ladders as well as water towers took advantage of the latest technology too. Improvements appeared from the end of the nineteenthth century through to World War II and beyond. Abraham Wivell, an English fire fighter, is credited with the first fly ladders, precursor to today's multilevel fire ladders. Wivell's 1830 creation comprised a main ladder tall enough to reach the top window of two-story buildings, a second extension ladder which could be pulled out to reach higher windows with various ropes and pulleys, and a third ladder which could extend to hook onto an even higher level. The whole was mounted on a spring wagon with large wheels and included a canvas attachment secured to the ladder as an escape chute, so that panicked fire victims could be rescued faster.

Wivell's ladders inspired more and more hook-and-ladder wagons, designed to be fast and versatile, able to reach a building with tremendous speed and to get in close to the flames. These "Escapes," as they were known in Europe, led to the eventual popularity of the turntable ladder in England and Europe as well as the aerial ladder in the United States. Unfortunately, cities were growing vertically as well as horizontally. The fly ladders of Wivell's day were not going to last very long, as buildings outgrew their reach. It required more than extensions – although that idea was not ignored – to solve the problem. The Scott-Uda aerial late in the nineteenth century, for example, used eight separate extensions with a variety of weights and balances for stability. Based on an Italian design, these aerials sold for a while but resulted in tragic circumstances – several men were killed during a public demonstration when the aerial collapsed. The model was never manufactured again and the Scott-Uda company faded from memory. A taller, safer model was needed. The first and most lasting developments were the turntable ladder,

patented in the United States in 1868 by Daniel Hayes, reaching heights of 85 feet, cranked into position by a vast worm gear and the strength of fire fighters, and aerial ladders such as the "Babcock" model introduced in the United States in 1886, again operated by a worm gear and hand cranks.

The history of these ladders is speckled with subtle innovation, including mechanical ladders, pneumatic and hydraulic models, ladders made of lighter and sturdier materials, and a popular caged model for added security. Motorization sparked the competitive spirit among ladder builders. The Merryweather motorized escape van of 1903, which offered a 50-foot escape ladder as well as a hose reel and a 60-gallon extinguisher, and the Merryweather turntable ladder of 1908, both represented some of the earliest work being done with motorized ladder apparatus in Europe at the turn of the century.

In the United States, Seagrave introduced its first aerial truck in 1902, with a 75-foot ladder. The first motorized aerial ladders came to New York in 1912, ordered from the Webb Motor Company. "Motorized" does not only mean gasoline-powered: many manufacturers experimented with both gas and electricity in their ladders. The Webb model used both gas and electricity to mount the ladder. One 1916 model by American LaFrance used air compression while another sat on a series of self-raising springs. By the end of the first quarter of the twentieth century, aerials and turntables had raised expectations in ladder technology. They offered flexibility,

FAR LEFT: **Wheeled escapes such as this Shand, Mason & Co. model were very popular in the UK, and stemmed from Abraham Wivell's models of the early nineteenth century.**

LEFT: **A 1919 American LaFrance water tower reaching 55 feet in vertical height.**

mobility, and greater safety to fire fighters and fire victims alike. Motorized water towers became equally important as ground fire fighting proved increasingly ineffective against taller buildings – the highest point of a fire was likely to spread to other buildings. It was no easy challenge to overcome. Manufacturers originally attached water hoses to ladders and raised water to a reasonable height, the whole secured to fire engine chassis for stability. Early attempts included the Skinner "hose elevator" of 1869, basically a raised hose, and the Greenleaf and Logans water tower of 1876. This model was secured to the floor of a wagon and incorporated varying lengths of pipe and a nozzle tip (it was raised by a crank-and-cable system and was quite effective). Though somewhat crude, they attracted the attention of local fire departments – each was sold and put into practical use.

George Hale of Kansas City, a true innovator in water tower design, produced a hydraulic system that made most water towers seem outdated. The hydraulics used a chemical combination of soda and acid, generating the gases needed to raise the tower. The system was so effective that Hale sold almost 40 of these towers by 1892.

The Hale design was eventually surpassed by the Fire Extinguisher Manufacturing Company of Chicago. The company bought the rights to the Greenleaf and Logan prior to the launch of Hale's invention. Faced with a sound beating in the market, the company created its own hydraulic model called the "Champion." It could pump both vertically and at an angle and was placed on a turntable for greater flexibility. The Champion represented the greatest competition to Hale's innovations. Hale sold the rights for his patent to the Fire Extinguisher Manufacturing Company and got out just in time. The Fire Extinguisher Manufacturing Company went on to become one of the only such manufacturers in North America. By the time it was amalgamated into the America-LaFrance Fire Engine Company in 1904, the Fire Extinguisher Manufacturing Company had become the unchallenged leader in water tower construction.

Motorization eventually made independent water towers redundant as aerials could serve the same function more effectively. American LaFrance, the new name given the Fire Extinguisher Manufacturing Company, and inheritor of the Hale, Greenleaf, and Logan designs, brought out its first factory-motorized vehicle in 1914. Seagrave was among the earliest competitors to American LaFrance – the first Seagrave motorized tower was delivered to St. Paul, Minnesota in 1917 and remained in use for the next seven decades. Most fire engine manufacturers introduced either an aerial or a water tower to its roster, a testament to the equipment's importance. The technology evolved to combine aerial with water tower innovations and do the job of both in a single unit. This drive toward streamlined machines and complete systems was essential as the clouds of war hovered once again and the need for even greater fire fighting equipment became a reality.

The Fires of War

5

INGENUITY WAS ALL-IMPORTANT DURING THE WAR

YEARS IN EUROPE. RATIONING MADE IT DIFFICULT

TO PRODUCE EFFECTIVE FIRE APPARATUS EVEN AS

DEMAND FOR THE ENGINES GREW. FOLLOWING

THE WAR, A NEW SPIRIT OF ENTHUSIASM AND

CREATIVITY, COUPLED WITH NEW IDEAS IN FIRE

FIGHTING TECHNOLOGY, PRODUCED SOME OF THE

MOST IMPRESSIVE EQUIPMENT SEEN TO DATE. NEW

STRATEGIES FOR FIRE FIGHTING BECAME ESSENTIAL

AS FIGHTING A FIRE BECAME MORE A MATTER OF

UNDERSTANDING IT THAN DROWNING IT IN WATER.

ILTON POINT HOSE COMPANY

LEFT: **Mack was one of the most popular models in the thick of battle during both first and second World Wars.**

PREVIOUS SPREAD: **The members of the Chiswick works (UK) were trained to face some of the hardest war-time conditions.**

Fighting the Good Fight

The evolution of fire fighting apparatus has always been radically affected by war rationing. War-ravaged cities and incendiary bombs made innovation necessary but difficult. The fight for independence in the United States for example severed ties for many manufacturers and buyers in America and England. Fire engine manufacturers began to bloom on American soil following the Revolution while, in England, innovation slowed and buyers became suspicious of new developments. During the Revolution, battles and fires raged but fire fighters were in short supply. Many were busy fighting the Revolution themselves, even though they were exempt from service. During the battles for New York, British soldiers attempted to fight fires that burned through the city but it was slow going. There were no bells left to call for help as fleeing American soldiers had taken them for munitions and any fire fighting equipment left was scuttled.

Fire fighting apparatus was in tremendous demand throughout Europe during World War I and some of the finest machinery came from the United States. Mack model ACs were so effective that many were bought after the war for use as fire fighting equipment. Practical innovation was seen on all sides of the war. Engines and motorized vehicles became a significant part of the fight. Everything from the aforementioned Mack AC trucks to hand-drawn engines by the Americans Obenchain and Boyer saw combat, along with Merryweather & Sons' motorcycle combination appliances carrying manual pumps to the scene of fire for quick and effective aid.

The need for practical solutions to seemingly insurmountable problems is to be expected in any crisis. What is not to be expected is the degree to which people can respond to adversity, as was the case during World War II in Britain. For fire engines and fire fighting techniques, World Wars I and II were a uniquely British experience. The years leading up to and during the conflicts produced some of the most interesting fire fighting apparatus in the U.K.

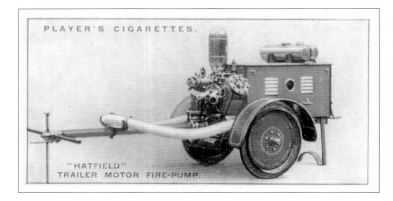

BELOW & RIGHT: **The 'Hatfield'**
trailer pump by Merryweather was
introduced in 1921 for rural areas
(35hp engine). This was a precursor
to the trailer employed during WWII
in England.

PLAYER'S CIGARETTES.

"HATFIELD"
TRAILER MOTOR FIRE-PUMP.

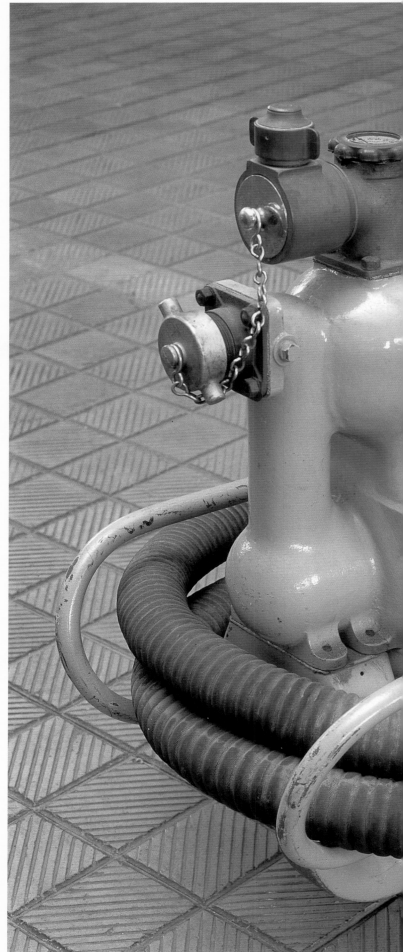

THE LESSONS OF WAR IN BRITAIN

World War I was a learning process for fire fighters in Britain and the British public at large. The need for professional fire fighters was extreme, prompting their exemption from military service. Their work was eventually considered as important as that of the average soldier, but it took some time to learn this lesson. The first air raids of World War I, via enemy Zeppelins and planes in both daytime and night raids, were often ineffective and too spread out to cause any long-term damage. They instilled enough fear and general concern, however, that fire fighters who had already been conscripted were brought home to continue their fire fighting duties. By this time the brigades had become sorely understaffed, almost one-third of the London Fire brigade alone having already enlisted and been sent off to war.

The war generated a growth in respect and admiration held for the fire fighting service by the public. Fire brigades had held a romantic grip on the imagination of the people for years, as horse-drawn pumpers and steamers bustled down crowded lanes to fight fires. Their reputation was marred somewhat by the rough side of fire fighting as infighting among volunteers and insurance brigades of the eighteenth and nineteenth centuries had shown. The World War I prompted a significant change in the public perception of the average fireman both in Britain and abroad.

The war also led to the eventual harmonizing of both fire fighters and fire fighting equipment in Britain. When the government and city councils passed a law requiring every fire brigade make itself available during a fire, they discovered a problem – the Greater London area had some twenty different hose couplings in use alone! The differences in the equipment being used mirrored the differences between brigades and prompted calls for a fire fighters' union. A fire fighters' union was finally and somewhat reluctantly established by the end of the war – the London Fire Brigade Representative Body, later known as the Firemen's Trade Union. Given the war-weary fire fighters' willingness to persevere in the face of such obstacles, the public perception of the fire fighter eventually became one of unreserved respect. It would take World War II, however, to iron out all the problems generated by World War I.

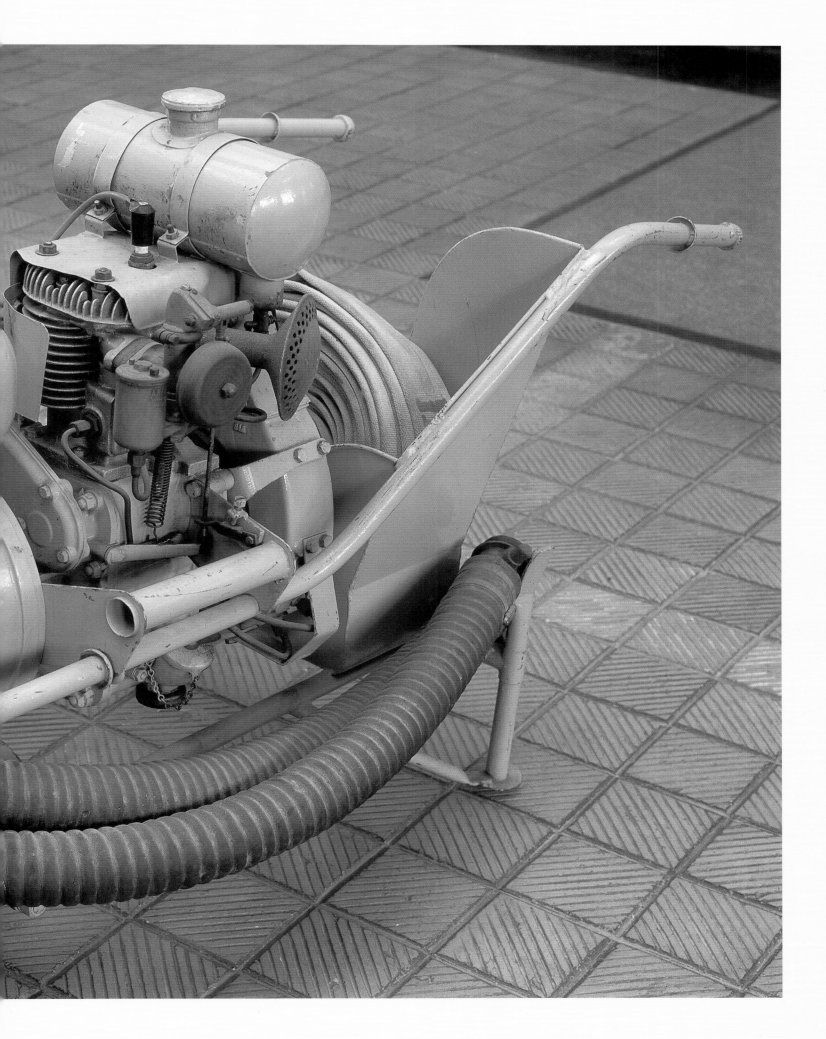

THE YEARS BEFORE WORLD WAR II

War brought out the best in the Britain's fire brigades as well as the worst. It brought brigades and firemen together in solidarity, taught valuable lessons to fire fighters, governing bodies, and the public as a whole, and exposed some of the country's fire fighting weaknesses. This was even more obvious during World War II, not only among the British but also among American and European fire brigades.

Britain prepared itself for the possibility of war from the outset through legislation and the assistance of volunteers. The Fire Brigades Division of the British Home Office pressed for greater fire prevention measures among localities by 1937, backed by the Air Raid Precautions Act, which offered subsidized funding for fire engine apparatus bought for this purpose. Recruiting was under way for the Auxiliary Fire Service (AFS) by 1938 and reserves were built up with an eye toward future possible air raids, a lesson carried over from World War I. It was hoped the ranks of the AFS would grow substantially from the start but this mark was not reached for the first year or so.

The British Home Office commissioned a standardized emergency trailer pump in response to public concerns about air raids. The trailer was considered essential under air-raid conditions and was eventually manufactured in the tens of thousands for free distribution across England. Their initial reputation probably stemmed from their inexpensive production costs and not their proven efficiency. They were brought to the scene of a fire in a truck or wagon – even taxis were used. Many brigades outfitted their machinery to tow pumps as well as fitting cars and any other available vehicles. The American-made Ford VB and 7V models joined the prewar efforts carrying pumping units, trailer pumps, and other emergency fire appliances before and during the war.

The commissioned trailer pumps were similar in design to earlier very popular engines like the Tilling Stevens models and the Dennis designs of the 1920s, but were even more practical. The first pumps were self-contained models mounted on trucks, but the list grew (from the smallest, which generated as low as 140 gpm, to the largest, which put out as much as 500 gpm). Different manufacturers got involved including many of the most popular names in England, such as Dennis, Scammel, and Worthington Simpson. One of the most famous was the "Coventry Climax," named after its manufacturing company and reputed to be the top of the lot – approximately 25,000 of these models alone were produced leading up to the war.

By 1939 the AFS had grown to include not only male reserves but women and teenagers as well. The equipment was that of prewar years, Dennis and Merryweather appliances still being used, but World War II was only just beginning.

BELOW: **1910 motorcycle and side car carrying a manual fire pump, built by Merryweather; motorcycles with pumps were able to deal with the bombed-out streets in war-time London.**

LEFT & RIGHT: **A 'Coventry Climax' trailer pump, produced by the thousands during WWII, the practical solution to air raid fires.**

OVERLEAF: **The Chiswick works (UK) in 1941 put their engines to the test.**

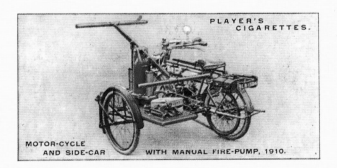

PLAYER'S CIGARETTES.

MOTOR-CYCLE AND SIDE-CAR   WITH MANUAL FIRE-PUMP, 1910.

WAR BREAKS OUT

From the moment Chamberlain finished his radio broadcast declaring war, fire brigades and the AFS braced themselves for air raids. The first anticipated attacks did not arrive as expected, however, allowing time for training and improvements. The extra training paid off but not before the first air raids of 1940. Before the bombs fell, the equipment was maintained and brought out for training missions in preparation. The motor on every engine was started each quarter hour during the cold winter of 1939 because there was no antifreeze. Water tanks froze frequently and had to be monitored closely. Public patience for the AFS and the bloated fire fighting forces was growing thin as days became weeks and no bombs fell. The AFS thinned as volunteers withdrew, prompting legislation to prevent any further departures. Then, just as everyone was beginning to think it couldn't happen, it did.

Many of the volunteers in the AFS had never come into contact with a fierce fire fighting situation by the time the first air raid struck – it was very much a baptism by fire. The members of the AFS fought the fires head on and took no shelter under attack. The earliest raids brought out over 100 pumps and multiple fireboats along the Thames, and millions of gallons of water were spread across the city, faced with an ugly urgency – night-time fires led each successive wave of bombers along their path.

Urgency led to the creation of the National Fire Service (NFS) in 1941, dividing the country into separate fire areas controlled by one central command. The air raids of 1941 spread outside of London before the creation of the NFS and many regions needed help from nearby fire brigades. Incompatible equipment and lack of consistency among fire fighting techniques created an even greater challenge. It didn't matter if a Dennis engine was powerful if it used a different hose coupling from that of its neighbors. A process of unification that had been sparked during World War I finally took hold among fire brigade communications, command structure, and equipment. By the end of World War II, Britain had one of the most organized fire fighting services in the world, due in a large part to the challenges faced and solved in war.

Though sometimes romanticized, wartime in England was never easy. In 1939, as the threat of air raids seemed more and more unlikely, AFS volunteers were sometimes considered idle hands. Under the barrage of enemy bombs, the AFS and the London Fire Brigade were hailed as heroes, facing some of the most dangerous circumstances imaginable. Those who had ridiculed their work, meanwhile, were hidden away in bomb shelters. By war's end, fire fighters, both professional and volunteer, were among the most highly decorated participants.

for air-raid attacks. APRVs were joined by the Auxiliary Towing Vehicle (ATV), essentially any vehicle used to tow the trailer pumps. Private cars and taxicabs were enlisted to transport equipment such as trailer pumps and ladders. Taxicabs took on this role with enthusiasm as a solution to their own lagging fares – they earned a government contract to act as ATVs. Taxi chassis were very sturdy and performed their roles dutifully, albeit for a fee.

Portable water dams were popular additions to this somewhat motley crew of equipment. These were large metal tanks that could carry up to 1,000 gallons of water. Trucks, buses, and carriages were also converted to carry these dams and do their part for the war effort.

Certain manufacturers stood out during the war and were among the most consistent suppliers. Many manufacturers were more than happy to take the government contract for trailer pumps. Aside from the Coventry Climax, Dennis, and Worthington-Simpson, there were also Beresford, Gwyne, Sulzer, and Harland models. Trailer pumps came in three sizes, small (140–175 gallon capacity), medium (230–320 gallon capacity) and large (500 gallon capacity), with powerful compact gasoline-powered engines operating either single or two-stage centrifugal pumps.

SUCCESS STORIES

Specialization during the war, such as Crossley's foam and CO_2 tender built specifically for Royal Air Force (RAF) landing strips, was another avenue of opportunity. The 1936 Crossley model included a 200-gallon water tank and four 80 lb cylinders for sulphuric acid and sodium bicarbonate (baking soda). The acid mixed with the sodium bicarbonate dissolved in the water to produce carbon dioxide of tremendous pressure, which forced the foam out. The foam smothered fires that broke out in crashed planes or burning engines – gasoline fires were a great challenge as water was not effective dousing this fuel source.

Chemical engines like Crossley's had been around since the steam era, introduced in 1872 by the Babcock Manufacturing Company of Chicago. The idea originated in France around 1864 for use in small household extinguishers but it wasn't until the Babcock engine that the idea bloomed. The chemical engine was an absolute necessity for the RAF, facing gasoline fires on their landing strips from crashed planes. The later Crossley FE1 model was built on a six-wheel chassis like the 1936 model and carried a 300-gallon tank plus a 28-gallon tank for liquid foam and four 60 lb cylinders for the CO_2. By concentrating its efforts on the needs of the wartime air forces in Britain, Crossley came through World War II both specialist in a unique and growing field and, manufacturer of some of the most effective machinery of the war.

The best-known fire engine manufacturer during the war and a company that remains at the forefront of fire apparatus development was the ever-popular Dennis Brothers. Dennis was established around the turn of the century and had shipped over 2,000 engines worldwide by the mid-1930s. The company's contribution to the war effort began before the war began with its 1935 Canteen Van, which distributed food and water to fire

FIRE ENGINES DURING THE WAR

Production of fire engines dropped during the war because of rationing and the appropriation of most heavy equipment for the war effort. Familiar names continued to produce engines but, in the UK at least, fire fighting apparatus earned no special status. The war was on everyone's mind. The efficiency and effectiveness of the machines built to help end the war or stem its effects were all that mattered.

One of the first war-specific developments in Britain, before even the aforementioned trailer pumps were commissioned, was the Air Raid Precaution Vehicle (APRV). These vehicles monitored the skies

THE HOME FIRES OF GERMANY

German manufacturers during the war produced of some excellent and efficient fire fighting equipment before and during the war. The Opel Blitz of 1939 was an all-wheel-drive vehicle with a front-mounted pump and fully enclosed body that could compete with anything available at the time. The Opel Blitz PWT had an additional 440-gallon tank for use as a street watering vehicle. Bombed-out streets proved as difficult for German fire fighters as for the English. In response, the German government authorized the manufacture of all-terrain fire engines after the war, supplied in large part by Mercedes-Benz. A Mercedes tactical vehicle called the "Unimog" was eventually outfitted for fire engine duties to deal with these unstable conditions and was so effective that some were purchased for use against brushfires in the States, although these appliances were too unstable for consistent use and were eventually discarded.

Though now known worldwide for their automotive achievements, Mercedes-Benz supplied fully enclosed Magirus-bodies cabs in 1941 of the highest quality. Other suppliers in Germany included Daimler, Klockner-Humbolt-Duetz, Opel, Hagen, Metz, Koebe, and Rosenbauer also contributed.

fighters. The Dennis hose engine of 1936 distributed over a mile and a half of hose via two hose reels as the vehicle moved forward at 15 miles per hour, yet another prewar innovation. The Dennis Ace of 1938 incorporated almost 2,000 feet of hose, a telescoping ladder, and foam-distribution equipment. The Dennis Big 4 and Big 6 machines of the 1930s were also very popular leading up to and including World War II. Dennis maintained its status throughout the war as it always has, with determination and a practical approach to the public need.

Fire apparatus manufacturers like Dennis faced the same challenges as most in Britain during the war. Few enjoyed consistent purchases and those who accepted government contracts faced rationing and shortages. Practicality was the key. Six-wheeled apparatus such as Crossley's foam tenders illustrate this pragmatic approach perfectly. Six-wheeled chassis were built out of necessity at the turn of the century in the face of rural conditions and unpaved roads. These six-wheelers were similar to the military units used during World War I and were extremely useful in the bomb-blasted streets of England. The extra axle provided additional support and the added length offered more room. Prestigious manufacturers like Leyland and Thornycroft-Simonis jumped on the six-wheeled chassis bandwagon and the children of these creations can still be seen today in the most rough and rural areas of the United States and Australia.

Innovations like the foam tender, the six-wheeled chassis, even the trailer pump seem insignificant next to the challenge of war itself but the dilemma of their builders was faced by everyone in the country – very few materials for manufacturing and even fewer people to do the building. The fact that production continued is a testament to those who met the challenge and persevered.

AMERICA DURING THE WAR

The war became a reality for most Americans on December 7, 1941 with the attack on Pearl Harbor. From that moment, America was committed to ending the war. The United States was still struggling with the after effects of the Great Depression during the mid- to late-1930s. With the onset of war, rationing was imposed on American manufacturers. Production ground to a slow trickle, saved only by the fact that fire fighting appliances were a wartime necessity – although each order required independent verification. Authorized wartime orders for fire apparatus and government contracts were all that kept many manufacturers going during these lean years. Machines were built without any chrome trimming as metal was rationed for munitions. The government could also seize any machinery it deemed essential, which meant that a city could find itself without apparatus at all.

Three companies managed to come through the fire fairly well due to military contracts: the Hale Fire Pump Company, Seagrave, and Mack. All three supplied the military and subsidized their income with the production of appliances for use in the United States. Hale produced tens of thousands of trailer pumps during the war in response to the demand from Blitz-ravaged England. Seagrave's naval contracts kept it afloat while the reliability and strength of Mack trucks kept them in high demand throughout the conflict.

Without exception, fire engine manufacturers were thrilled with the end of the war, although recovery would take some time. When the dust settled, American manufacturers could look forward to some of the greatest years in their collective history.

ABOVE: **A 1940 Mack quad, at its best in war and peace.**

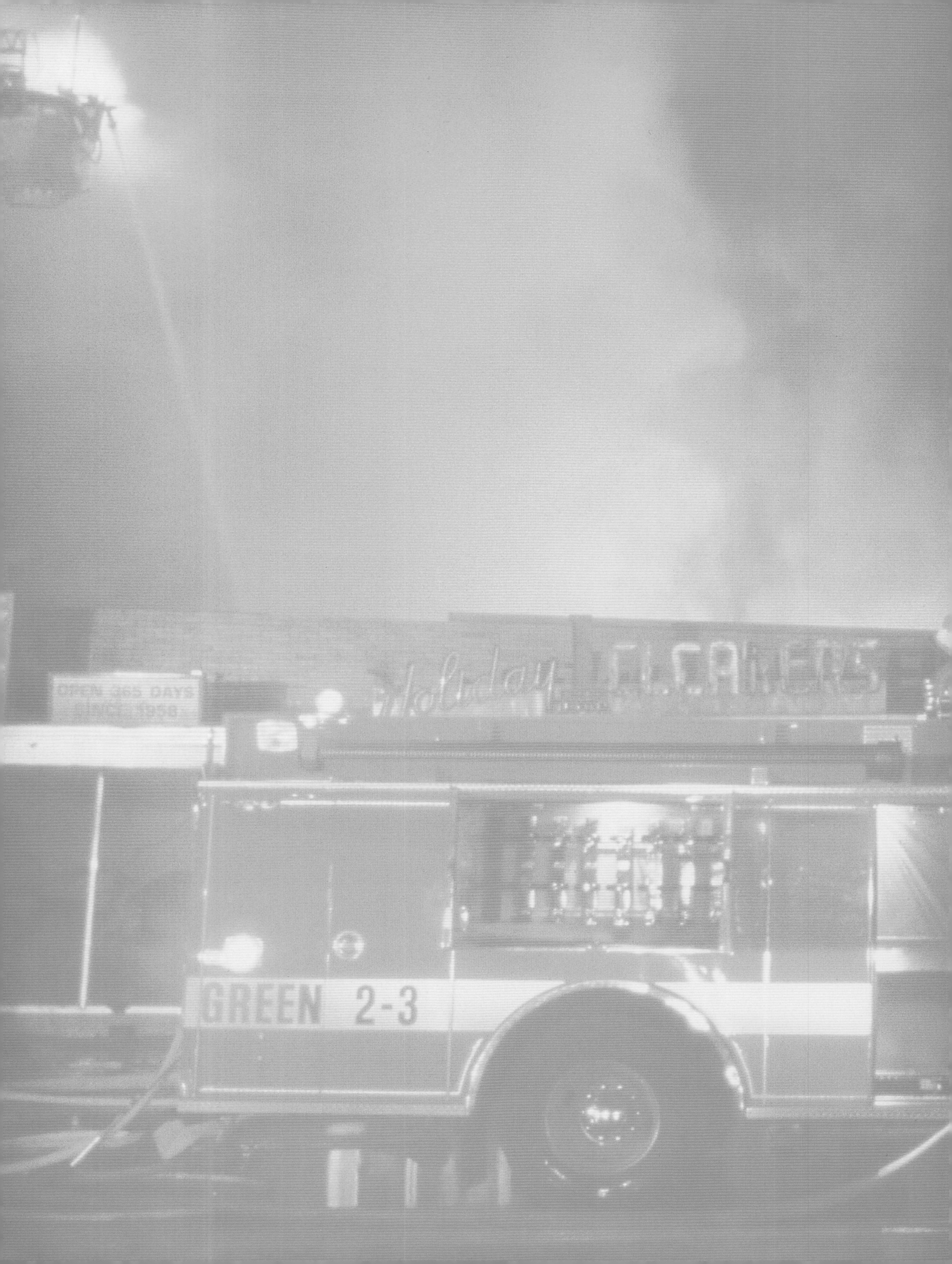
OPEN 365 DAYS
SINCE 1958
Holiday CLEANERS
GREEN 2-3

The Fire Next Time

THE YEARS FOLLOWING THE WAR'S END WERE AN ECONOMIC BOOM FOR THE UNITED STATES BUT A CHALLENGE FOR ENGLAND'S UNSTABLE AND WAR-TORN LAND. FROM 1945 ONWARD INNOVATIONS THAT SURFACED WERE ADOPTED QUICKLY. THE POST-WAR BOOM OF THE 1940S AND 1950S MEANT EXPANSION FOR AMERICAN MANUFACTURERS AND OPENED DOORS FOR NEWCOMERS. THE PAST 40 YEARS HAVE PRODUCED A MORE STRATEGY- AND PEOPLE-CONSCIOUS APPROACH TO FIRE FIGHTING AS WELL AS TECHNOLOGY TO COMBAT NEW THREATS ON THE HORIZON. FOR THE MILLENNIUM THAT LIES AHEAD, THE POSSIBILITIES ARE LIMITLESS.

The Future of Fire Fighting

T he war generated new ideas, innovations in fire containment, and revised strategies for combating a blaze, although water remained the most consistently effective fire fighting technology. There was a renewed spirit among manufacturers as well, particularly in America, and the decades following the war were the most productive. Greater quality of design was the focus, along with increased reliability, durability, and an effort to include the latest technology in the fire fighting realm.

AFTER THE WAR

The first burst happened in the United States, where communities that were unable to afford fire fighting apparatus due to the Great Depression and World War II suddenly felt a pressing need for this equipment and had the money to buy it. Suburban housing began to spread as soldiers returned home and started families. The GI Bill offered returning soldiers the chance for higher education and they took advantage of the opportunity in droves. America was in full swing while the pieces were still being picked up in Britain and the rest of Europe – it would be a decade before the field would become more or less level once again.

This post-war era of prosperity gave fire engine manufacturers in the United States their greatest audience. Communities needed fire fighting equipment and suppliers did their best to keep up, although most did not get back into full operation until 1948. This created a backlog of orders throughout North America – some communities required the replacement of whole fleets of equipment. Los Angeles alone, for example, bought eighteen triple combinations from Peter Pirsch & Sons and an American LaFrance 100-foot aerial powered by a V12 engine in 1945, as well as a Seagrave 100-foot aerial in 1946.

In Britain and the rest of Europe where the war and the after effects of the Depression were felt for much longer, manufacturers relied on proven designs and few innovations. Appliances that were effective during the war years were adapted to public use and kept in service for years. Very little creativity lingered in the spirit of those manufacturers that had survived the war. The Dennis Brothers were still leaders in the UK and produced open cab designs, switching to Rolls-Royce engines and buying the British franchise for the German Metz turntable-ladder mounted on a forward control chassis. Other European manufacturers at the time included Kerr, Foamite, Morris, Angus, Sun, Hampshire, Haydon, Whitson, Carmicheal, Campion, Pyrene, Miles, and Arlington, but none could compete with the enviable position afforded America at this stage in fire engine history.

ABOVE: **A 1946 Ford/John Bean, high pressure pump.**

LEFT: **A Dennis Metz 100 foot UK rig from the Glasgow Fire Service.**

FAR LEFT: **Dennis was and is one of the more popular brand names in fire engines in England.**

PREVIOUS SPREAD: **The modern fire engine has developed fairly quickly over the last century and improved fire-fighting greatly.**

POST-WAR INNOVATION

Despite the money thrown at many manufacturers – or perhaps because of it – innovation during these golden years was more subdued. Consistency, stability, and quantity were vital as manufacturers in the States sought to keep up with the demand and reap the rewards. Creativity suffered in a sink-or-swim atmosphere as no manufacturer was willing to invest time or money in an idea that wasn't immediately effective. Given this sort of environment, post-war innovations that took on long-term significance merit particular attention.

One was the creation of the cab-forward design featured in today's heavier equipment. The cab of the engine was placed in front of the engine itself, providing greater visibility for the driver as well as greater room for passenger fire fighters, making them a practical option for many burgeoning fire departments. The cab-forward design also meant tighter turns, a shorter body design, and

ABOVE: **An American LaFrance 1940 model.**

RIGHT: **A 1954 American LaFrance with open cab and 85 feet of ladder.**

FAR RIGHT: **This 1959 American LaFrance model carries 75 feet of ladder.**

PREVIOUS PAGE: **A 1947 American LaFrance (700 Series) with 750 gpm.**

BELOW: **A 1968 Four Wheel Drive Pierce with 85 foot snorkel.**

RIGHT: **The offspring of the Quinn snorkel in action, providing greater mobility for better fire fighting.**

a lower engine for greater stability. American LaFrance deserves most of the credit for this innovation, producing its first such model, the now famous Series 700, in 1945. American LaFrance placed the cab ahead of both the American LaFrance V12 engine and the front axle and, in a rare show of acceptance, the new cab forward design was applauded throughout North America and eventually in Europe. The design was used by manufacturers of fire engines and heavy machinery in both countries, as well as in buses and some haulers. The flat, snub-nosed look of these engines is a familiar sight today but at the time the flat-faced engine was regarded with awe. After years of bland machines cleared of all chrome plating, this engine was greeted with enthusiasm in fire fighting quarters. Praise for these machines was so great that by 1947, American LaFrance had fitted its factories to build the Series 700 exclusively.

Another significant development during the war was the high-pressure fog engine, created by the US Navy. It delivered a sheet of fog similar in theory to conventional smoke machines but under higher pressure. The dense fog lowered temperatures at a scene and broke the path of the fire. These engines were fast, effective, and ideal for shipboard fires, used very little water, and required only a small pump to operate. Producing fog generates little actual weight, making the engine light and less damaging than a jet of water. Unlike water, fog provides a shield for fire fighters, a wall blocking the unbearable heat of a raging fire. When water is used against a fire it can produce a backlash of boiling steam as

dangerous as the flames themselves. Unlike chemical systems, fog doesn't smother a fire but attacks the temperature, reducing the chance of its sparking again. The Food Machinery Corporation produced high pressure machinery during the 1940s and into the 1950s and offered a standard high-pressure fog unit as well as a high-pressure fog gun. A popular model was produced by Autocar in 1950, ready for operation in less than 15 seconds and requiring no hydrant! The high-pressure fog system was a unique and practical creation but was never adapted on the widespread basis it deserved. It did enjoy a niche market which could not be beaten.

Perhaps the most famous of all post-war adaptations is the Quinn Snorkel, the famous offspring of the cherrypicker. The Quinn Snorkel was based on Chicago fire commissioner Robert Quinn's inspired idea to adapt the aforementioned cherrypicker. It provided greater access to taller buildings, more efficiency for rescues, and greater volumes of water than ever projected by any water tower. The snorkel used similar technology to that of its parent, a mobile bucket at the top of an extended mechanized boom with an added length of hose and nozzle. The snorkel could reach great heights with more certainty than anything on the market and was folded when not in use for easy, aerodynamic transportation. The idea was celebrated throughout the country and snorkels spread across North America and beyond. You are likely to see this innovative adaptation in use in America or Europe, as well as in Australia and Japan.

BELOW & RIGHT: **Dennis (UK) produced a distinctive range of machines, from self-contained enclosed cabs to ladder engines of varying heights, used throughout England.**

AFTER THE FIRE

Europe suffered economically for years after the war, requiring massive rebuilding schemes that drained the coffers of most post-war reparations, leaving communities struggling to get by. Europe didn't enjoy the sort of money available in America and manufacturers struggled with the after shocks of war. Some rural areas of Europe used manual pumpers and steamers well into the 1960s as a consequence, while others relied on the goodwill of local neighbors.

Dennis Brothers sold enough machines during the war to maintain its place at the forefront of British fire apparatus manufacturing, buying out the British franchise for the Metz turntable ladder and incorporating it into the standard design in the process. The company shifted from its own engines to sturdier Rolls-Royce engines, as in the Dennis F2 series in which a Rolls-Royce B80 eight-cylinder engine provided more than enough power for the sturdy apparatus. The F2 was a narrow machine, only slightly wider than the Dennis F1, one of its first post-war productions. It was ideal for rural service. The Dennis F series did well for the company and it was quick to offer a great variety. The Dennis F6 model tanker even had an optional sprinkler system attachment for street cleaning. Dennis offered everything from foam engines to water tenders, from combination units to ladder-specific carriers. Of all the producers in England after the war, Dennis fought to survive and took advantage of its position wholeheartedly.

Leyland also expected to come out of the war at the forefront of British fire engine manufacturing but the company didn't recover as well as its rival. Dennis was able to buy the British right to the Metz ladder because its main competitor, Leyland, decided to concentrate on commercial vehicle production instead of fire engines. Leyland delivered its first engine in 1910 and was one of the best producers of fire engine appliances in Europe. When the company released a selection of its "Comet" as a variety of small fire appliances, the fire fighting world watched in anticipation but was given only crumbs. When Leyland brought out its "Firemaster" engine in 1958, it was perceived as the first in a new line but once again the anticipation never played out. The Firemaster was up to date, offering a typical Leyland coach chassis and an engine which included semi-automatic transmission set mid-chassis and a low-level pump for a cleaner floorboard. Leyland unfortunately produced only a handful of these machines after the war but the hope for a new face on the fire engine horizon in Europe remained strong.

ABOVE & LEFT: **Leyland, though not a strong contender today, was quite popular in its day throughout England.**

ABOVE RIGHT: **Merryweather, a powerful force on the English fire engine scene, remained strong well into this century. These models could be found as far north as Glasgow.**

Merryweather, another British manufacturer from whom great things were expected, was focused on finding the right chassis, for its engines – it had abandoned its previous Albion-Merryweather chassis. It settled on an AEC Regent Mark III chassis, and the Merryweather-AEC combination proved more than satisfactory to buyers – no surprise as the companies had proven themselves consistently over the years. The partnership continued into the 1960s, the 1963 Merryweather Marquis representing the perfect picture of this hybrid. This model was a high-performance machine with a diesel engine producing 130 horsepower. The Merryweather pump could jet approximately 1,000 gpm onto a fire. Over the course of its history, Merryweather had produced some of the best engines seen in Britain and abroad, exporting as far as Hong Kong (a spectacular Merryweather aerial ladder was delivered to the island in 1933). It remains one of the most highly praised producers of fire engine apparatus worldwide.

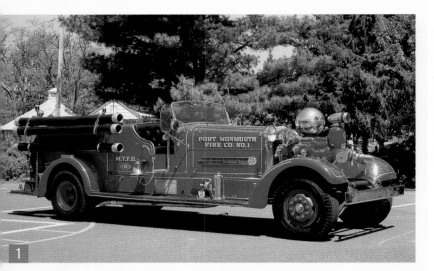

1

2

3

4

1: **Ahrens-Fox remained committed to quality. Pictured is a model from 1947.**

2: **This Ahrens-Fox from 1948 shows a quite dramatic shift in appearance from its previous incarnations.**

3: **By 1956, again incorporating a major change, Ahrens-Fix were offering cab-forward designs.**

4: **Compare this 1955 Ahrens-Fox with models which followed – both very different, but each retaining their unique style.**

5: **A 1952 Ahrens-Fox – the rounded grille and hood plus the inclined windshield distinguished the engine from the competition.**

Prospects in the United States were brighter but more competitive as many corporate rivalries and mega-mergers were going strong while others fell by the wayside. The saga of Ahrens-Fox is one story of struggle, success, and of failure, the proud manufacturer of some of the most impressive machines to come out of the American industry. Ahrens-Fox had a reputation for costly engines during the 1940s and 1950s and suffered when the public turned to more commercial and less expensive models after the Great Depression and the war – customized rigs were no longer in high demand, forcing Ahrens-Fox to become commercially conscious in order to survive. The company merged with the LeBlond-Schacht Truck Company in 1936 to stem financial complications, incorporating LeBlond-Schacht chassis into its designs. Despite favorable responses after the New York World's Fair, Ahrens-Fox found itself in trouble by 1939, forcing the company into liquidation and massive reorganization. Financial mishaps were to recur regularly during its final years as it shifted between owners – from Walter Walenhorst to the Cleveland Automatic Machine Company, from General Truck

Sales Incorporated to C.D. Beck and, finally, Mack Trucks. Market demand even forced the company to give up many of its distinguishing markers, such as its front-mounted piston pump, and convert to cheaper centrifugal pumps. With bittersweet irony, the company managed to last long enough to celebrate its centenary in 1952 but had little to celebrate. The company was revived in 1953 by the C.D. Beck Company of Ohio and surprised everyone by launching a new line of cab-forward models in direct competition with American LaFrance, the only other producer of these models. Unfortunately, the C.D. Beck Company was bought out by Mack Trucks in 1956, who designated the new model the "Mack C" Series and effectively eliminated the Ahrens-Fox name, at least for a time. The Ahrens-Fox Fire Engine Company did not die. It was purchased by in 1957 Curt Nepper, who celebrated Ahrens-Fox fire apparatus until his death in 1993, offering advice and swapping stories with fellow Ahrens-Fox enthusiasts. The Ahrens-Fox name has since been in the possession of the Menke family who continue the tradition with love and care.

1: Mack, ever strong on the marketplace, offered this model 85LS in 1949.

2: 1946 model Mack.

3: Note the bumper in this 1943 Mack model 75 – the chrome was added only after the war.

4: 1948 Mack 85LS, producing 750 gpm.

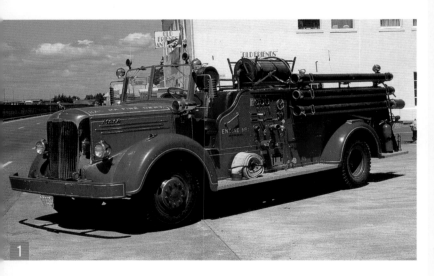

Mack Trucks came through the war fairly unscathed. The wartime experience of these sturdy vehicles gave them a greater reputation for durability than ever – over 10,000 of the very rugged Mack L Series trucks alone were distributed between 1935 and the early 1950s, distinguished by an exceptional dual-ignition engine with excellent pumping performance. Their post-war success reflected their reputation. Mack manufactured fire apparatus throughout the war and picked up in peacetime where it had left off, offering various designs of various sizes, all with the distinctive Mack Truck durability and quality. Double and triple combination pumpers were available, ranging from 500 to 1,000 gpm, and the buyers were many.

The 1953–54 introduction of the Mack B Model truck was the proverbial next step in the company's fire engine history. The B Model represented a new line that proved one of the most popular truck and fire engine designs of all time, easily identified by its large, plated radiator grille. This design became the Mack trademark look. It ranged from 500 to over 1,200 gpm and came in a variety of sizes and styles until around 1967, by which time it had become one of the most the prevalent models in the United States. It incorporated the powerful Mack Thermodyne six-cylinder diesel engine and could generate tremendous power, bearing the bulldog ornament for which the Mack name had become famous worldwide. Mack always recognized a good thing when it saw one and introduced the Mack C Series around 1956 – formerly the Ahrens-Fox cab-forward design. The Mack C was snatched up eagerly by buyers who also recognized a good thing when it came along. The C Series Mack truck was so popular that these models were eventually used to upgrade entire departments – almost 100 rigs were delivered to one department by 1959 alone.

5

6

7

8

9

5: **1951 Mack 505-A.**

6: **A 1954 Mack LT 21-A, including a Hall Scott gas engine.**

7: **This 1954 Mack LT 21-A was the only one of its kind ever built.**

8: **1953 Mack LS producing 750 gpm.**

9: **This 1950 Mack-Maxim combination included a 75 foot ladder.**

10: **1957 Mack quad.**

ENGINE 6

10

1: **1941 American LaFrance quad (625 gpm).**

2: **1941 American LaFrance 500 series.**

3: **1944 American LaFrance 600 series (750 gpm).**

Other familiar names took advantage of the post-war boom, although not to the same degree as Mack. American LaFrance introduced the cab-forward design to the world and did well into the 1950s, though the cost to retool the factories and manufacture the new design required additional financial backing. In keeping with its long history of amalgamation, going back to 1832, American LaFrance was sold to the Serling Precision Corporation in 1956 but maintained business as usual for years after. The 1950s were quite good to the company as it produced airfield tenders for the U.S. Air Force between 1950 and 1953. It also chased smaller markets by introducing smaller pumps similar to their original Series 700 cab-forward design which produced up to 1,000 gpm. The later Series 800 and 900 marked a turn for the better for the company and provided greater mobility coupled with improved quality. The 1962 Series 900 could be equipped with an immense elevating platform, the engine's large and stable body the ideal base. American-Lafrance managed to maintain its hold on the public imagination over the years. It changed with the times and manufactured some of the best equipment for fighting fires during this post-war era.

4

5

6

4: By 1954, American LaFrance could offer a model producing 1,500 gpm.

5: American LaFrance from 1944 with 85 foot ladder.

6: 1949 model American LaFrance.

7: 1946 American LaFrance with 65 foot ladder.

8: 1953 American LaFrance 700 series (750 gpm).

9: 1946 model American LaFrance.

1: **1947 Seagrave model 80 (750 gpm) – after the war, Seagrave worked hard to keep up and survived through amalgamation.**

2: **1949 model Seagrave.**

3: **A 1955 Seagrave model 900B (1,000 gpm).**

4: **A 1957 model Seagrave.**

5: **1946 Seagrave model 66 (1,000 gpm).**

6: **A 1958 model Seagrave.**

Seagrave and Maxim, both famous prewar names, came out of the war in fairly good shape and joined forces to combat harsh economic realities (Seagrave bought Maxim in 1956 but both operated independently). Seagrave focused its attention on enclosed cabs during the 1940s while Maxim prepared a new line of engines before the war had even ended. Seagrave's 70th anniversary in 1951 launched the appropriately named "70th Anniversary Series," so popular it was still being manufactured almost 20 years later. Seagrave stayed with engine-forward designs during the 1940s and 1950s, offering a cab-forward design in 1959. Nonetheless, Seagrave remained loyal to engine forward design in most models. Maxim, meanwhile, returned to its speciality after the war: aerial ladders. Maxim offered a range of hydraulic aerials that were powerful, dependable, and available on every kind of chassis imaginable, their control panel located alongside the ladder itself. These were among the best of the 1950s and into the 1960s. In 1963, Seagrave (and Maxim) was purchased by the Four Wheel Drive Corporation (a division of General Motors) and renamed the Seagrave Fire Apparatus Division. The Seagrave name can still be seen on fire apparatus and heavy-rig chassis around the world, while the Maxim name, unfortunately, has faded.

7: **A 1965 Four Wheel Drive model.**

8: **A 1949 Maxim (750 gpm) – Maxim was bought by Seagrave and eventually faded from the picture.**

9: **A 1951 model Maxim.**

10: **Four Wheel Drive model from 1963.**

RIGHT: **A 1951 Dodge-Pirsch with 55 foot ladder.**

BOTTOM LEFT: **A 1950 model Pirsch.**

BOTTOM RIGHT: **A 1950 Howe-Defender 750 gpm engine with Duplex chassis.**

Peter Pirsch & Sons focused on enclosed sedan-style models after the war and continued building fire apparatus on any chassis the client wanted. This was often an ideal financial solution for smaller fire departments without the resource to buy larger, complete models. The Howe Fire Apparatus Company reestablished itself in the 1950s with a redesign of its "Defender" model originally launched in the 1930s. The 1953 version of the Defender was bigger, bolder, and a much stronger engine than anticipated. It ranged from 750 gpm to the standard 1,250 gpm version. The Defender was upgraded in 1960 as Howe joined the ranks of cab forward manufacturers, and included a roomy interior for five fire fighters. Like Pirsch, Howe used commercial chassis for its models, relying on familiar names like Ford, Chevy, and the ever-popular REO Speedwagon. Howe was eventually bought by the Oren Nash Fire Apparatus Company which was subsequently bought by the Grumman Emergency Products Corporation in 1976. By the early 1980s, the Howe name was no more although it had made one of the more impressive contributions to the industry with its raised pump control of 1967. Moving the standard pump control to the back of the cab meant greater visibility for the pump controller at a fire. This was significant, as the pump controller would need this visibility for maximum efficiency.

The years after the war and through the 1950s were a time for growth and a test of every company's skill, some facing the challenge of intense demand, others the possibility of imminent closure. These years introduced some of the most streamlined fire apparatus and well-built custom rigs ever seen.

THE SIXTIES AND BEYOND

The 1960s and 1970s were defined by social upheaval and racial tension worldwide. Fire fighters found themselves faced with angry mobs starting fires they did not want put out, inner-city conflicts where they were not welcome, and cutbacks that left them strained and short-staffed. The glory days of the noble fire fighter were ending.

It took a new type of Great Fire to provoke change among fire departments, particularly in the United States. The Watts riots of 1965, the Detroit riots after the assassination of Martin Luther King Jr. in 1968, even the L.A. riots of 1992 – each was a reminder that fire fighting brigades could do only so much. Fighting a fire during a riot presented enormous challenges as rioters blocked access and caused injuries needing immediate treatment. As one blaze was put out, rioters would set another building on fire, even as fire fighters were trying to help the injured and stop the spread of destruction.

Fire fighters had to develop new strategies and equipment to do their job. Walls were knocked down during the 1968 riots to create fire breaks so fire fighters could move on and keep the conflagration from spreading any further. Rescue rigs or squads were developed after the first sparks of civil unrest, which included everything from hydraulic jacks to first aid, from the "jaws of life" to open a crushed and mangled car to oxygen for victims of smoke inhalation. Fire fighters tried to do their job despite the added complications. They would see the fight through as always and their machines were refitted in order to keep up with the changing times.

The diesel engine was one refit that survived. Gas shortages and economic pressures forced manufacturers to develop more efficient and effective systems to put out fires. A concerted promotional effort by Mack Trucks during the 1960s contributed to the acceptance of diesel engines as an economically sound alternative to standard gas-powered engines. Once their Thermodyne engines

1: **1961 Mack model B85 (750 gpm).**

2: **1961 Mack model C95 (1,000 gpm).**

3: **1962 Mack (750 gpm).**

4: **1964 Mack, with a leap forward in power, offering 1,250 gpm.**

5: **One of the few open cab-forward designs by Mack from 1969.**

hit the market, the industry was convinced. Diesel engines were patented in 1892 by Rudolf Diesel, a German engineer, and proved more efficient for large-scale machinery, as well as a reliable, cheaper alternative – the perfect choice for fire engines. Companies like American LaFrance saw the potential and, having already switched from their own V12 engines to the cheaper and more effective Waukesha engine, the transformation to diesel was not far away. American LaFrance introduced its first diesel engine in 1965.

Diesel was not the only innovative solution of the 1960s and 1970s. When White Motors added a tilting cab to its cab-forward design for example, Mack jumped at the chance to open up its rigs for easier maintenance. Mack launched the N Series trucks in 1958, which incorporated this unique design feature. The tilting cab opened up the engine compartment for easy maintenance and convenient access. By the mid-1960s, the cab-tilt was a common feature on most cab forward designs.

RIGHT & BELOW: **The Mack Super Pumper Complex, one of the most extraordinary developments in fire engine history, as impressive now as when it was first introduced in the mid-1960s.**

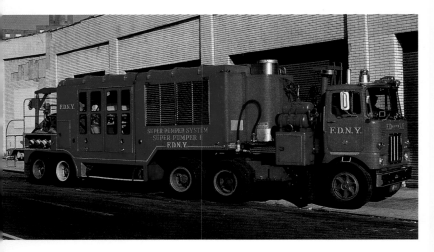

Subtle design alterations have been the dominant facet of fire engine development during the past 40 years as cities grew taller and public safety was paramount. Growing skylines required greater amounts of water than ever and more powerful engines to get it to a fire. The obvious solution was to build a bigger rig. The Mack "Super Pumper Complex" of 1965 was based on experiments conducted by the U.S. Navy and the British Admiralty and inspired by the fireboats which had been in service for years. Both military bodies wanted a lighter, more powerful engine for their ships. The Napier-Deltic engine was the perfect solution. Fireboat engines used a stronger hose and high-pressure nozzle to control the great volumes of water generated. These developments hadn't been incorporated in surface rigs due to a lack of chassis able to cope with their size and power.

Mack Trucks, at the peak of its popularity in the industry, decided to test its equipment against these engines in 1965. The result was the Super Pumper Complex. The Super Pumper was a fire fighting system made up of six parts (a Super Pumper, Super Tender, three Satellite Tenders, and a station wagon for the controller), bringing outstanding power against the types of fire that were anticipated among the skyscrapers of New York City. The idea was to produce a system of engines that could pump enough water to take on any fire. The Super Pumper was set up at the biggest water source available, or multiple sources if necessary, and provided a constant flow of water through four to eight hoses, to the Super Tender and its Satellite Tenders – an enormous fire fighting octopus. The Napier-Deltic engine provided enough power to distribute over 8,000 gpm, more than enough water for any fire. So much power is generated by these machines that the Super Tender had to have additional hydraulic support – without it the 10,000 gpm water cannon would have pushed the rig back in its tracks. The Super Pumper Complex was powerful enough to replace up to 10 rigs. When the original was retired in the early 1980s it was replaced with a total of six double pumpers.

The Super Pumper was the only all-encompassing innovation in post-war fire apparatus but it will not be the last. Since the introduction of the Super Pumper, developments have taken

advantage of available technology to greater effect. Stronger, more resistant insulated hoses, more efficient engines, stronger commercial chassis, and revised fire-fighting strategies have resulted in fewer fires and greater dedication to public safety.

The 1970s and 1980s brought more streamlined pumpers and attention to detail – all aspects of fire fighting were being considered. Combination pumpers were built that could generate upwards of 2,000 gpm in a range of strengths. Each new breakthrough or technological innovation, however, could just as easily complicate the life of the average fire fighter. Buildings grew even taller and pumpers and ladder trucks strained to reach that little bit higher. Specialized models sprang up as quickly as they disappeared, while commercial models became more powerful. The key to the last 20 years of fire engine development has been the search for the complete unit.

Specialization produced some of the more interesting engines and apparatus after the war such as air field crash vehicles. Depending on where you are, these machines are known as *airfield crash tenders, crash vehicles, crash tenders,* or *airport crash trucks.* No matter what you call them, these engines need speed, versatility, mobility, and power. A crashed plane is often an explosion waiting to happen as the fuel needs only a small spark to set the whole plane ablaze. It has become more and more apparent over the last 50 years that these specialized engines were not only necessary, they required strict attention to detail as they faced some of the potentially worst fires imaginable.

ABOVE LEFT & LEFT: **The Mack Super Pumper was a combination of tenders, pumpers and a controller. The high pressure hose on this satellite could hit 2,000 feet through a 4$\frac{1}{4}$ inch hose.**

ABOVE: **The 1973 Oshkosh (M-1000) foam tender was very effective fighting fires on the airfields.**

The need for these specialized engines first came up during the war as planes carried more fuel and were prone to exploding on impact. Familiar names took the lead and produced formidable machines to fight these flames. Dennis was among the first to launch an effective and powerful air field tender in Britain but Pyrene led the way in the most effective crash tenders, using commercial chassis from Dodge, Bedford, Scammell, and Thorneycroft. The crash tenders Pyrene produced incorporated everything that had been learned on the airfields during the war. The Bedford/Pyrene QL model of 1947 carved a path for every air crash vehicle to come. It was capable of all-wheel drive, carried 500 gallons of water along with a mass of chemical foam and CO_2 extinguishers – the QL was versatility itself. Pyrene recognized the need for mobility in its crash tenders and produced trailer-mounted fire engine monitors as a consequence, providing access to all sides of a crashed plane as well as greater proximity. Pyrene led the way in the 1970s with its tenders and this decade represented the peak of air-crash tender production.

The Pyrene Pathfinder, built in 1971, was state-of-the-art and streamlined the design of the air-crash tender.

Airplanes had grown larger and the damage that could be caused by their crashing had grown exponentially since the war. More fuel, more passengers, and more planes in the air meant ever greater risk. Pyrene was one of the first crash-tender manufacturers to recognize the need for greater flexibility at a scene. First, speed was vital against the burning liquid fuel surrounding the plane and the temperature had to be lowered as quickly as possible. Flooding the site with foam could bring a scene under control the fastest. Second, the engine had to get in close to get the passengers out and begin rescue operations. The Pathfinder was able to do all of this and more. Compared to its predecessors, it was almost five times as powerful, carrying 3,000 gallons of water, over 300 gallons of foam, and a Coventry Climax engine that could generate 1,500 gpm (compared to the 400 gpm of the Climax used in the Bedford/Pyrene QL).

BELOW: **Dennis (UK) is as powerful as its American counterparts.**

RIGHT: **Current fire engine technology, as in this contemporary Volvo machine, include hydraulic support legs for added stability.**

THE FUTURE OF FIRE FIGHTING

The engines being built today and the corporations building them are always looking to the future. Emergency One, Mack Trucks, Pierce, Volvo, Dennis, Leyland, and others continue to work on their designs, updating them for a future where the threat may no longer just be from fire. Today's truck provides protection for its crew, includes rescue equipment, and incorporates both "space-age" technology and race-car manufacturing standards.

Dennis remains one of the most respected and oldest names among fire engine manufacturers in Britain today. Its two signature models, the Dennis Rapier high performance vehicle and the Dennis Sabre general performance fire engine, can be seen barreling down the streets of London today, racing through the city's still-narrow roads through heavy traffic to respond to an alarm. Appropriately enough, given these conditions, Dennis's main concern in manufacturing these chassis is their handling.

The Rapier chassis has been redesigned with "space-frame" construction, offering less weight through lighter, more durable materials. Smaller wheels and tires and independent front-wheel suspension with "anti-dive geometry" are used to improve the truck's efficiency, similar to a racing car. Both the Rapier and the Sabre incorporate very powerful Cummins C260 Euro 2 six-cylinder turbocharged or intercooled diesel engines, a forward cab capable of 42° tilt, plus stainless-steel corrosion-resistant construction. In both cases, safety and performance are primary concerns and virtually every need of the fire fighter is considered, from seatbelts to shatterproof windows. Fire departments can choose from three different sizes (500 gpm, 750 gpm and 1000 gpm) providing fire fighters with power and maneuverability on the road. The Rapier, with its narrower wheelbase and shorter body, is designed for high performance under the strictest circumstances, perfect for the tight curves and urgency of a contemporary urban fire. The Sabre, meanwhile, is designed for general fire fighting, the meat and potatoes of the business. These machines have been built for every contingency and have proven their worth yet again. Dennis continues a proud tradition that has been ongoing for over 100 years.

LEFT & BELOW: **Today's engines must be prepared for any emergency, as demonstrated by the Volvo rescue vehicle and full ladder engines in use in the UK.**

Volvo Trucks, another familiar name among European fire vehicle manufacturers, produces some of the foremost chassis and engines available. It has a long tradition of producing the safest vehicles on the market through its strict testing regimen. More than half the fire services in the United Kingdom have used Volvo vehicles in active service. These versatile engines range from 6 to 10 liters and power Volvo's smallest to largest engines dependably, contributing to Volvo's overall solid reputation. Customers are offered a different range of chassis sizes and Volvo matches the engine, gearbox, and final drive to produce the best possible engine. As for safety, there are few companies in the world today that offer the sort of testing that Volvo incorporates into its manufacturing process. Its "Swedish Cab Impact Test" alone is unique in the industry and ensures Volvo cabs can withstand any beating. Volvo also offers full support and maintenance to its clients 24 hours a day, 365 days a year. Today's fire engine manufacturers are no longer just builders of fire engines: they provide comfort, economy and safety to their customers at all times.

Despite advances in European manufacturing, the largest fire engine builders worldwide are still found in the United States. Pierce Manufacturing, founded in 1913 to build trucks and bus chassis, continues to provide clients with a wide variety of vehicles of the highest quality. The company is conscious of both the small fire department as well as large, its product line including smaller models such as its "Suburban" and its "Responder" as an inexpensive option for many middle- and small-size fire departments. Pierce offers everything from small and speedy "Dash" models to its industrial strength "Quantum," built for "maximum room, comfort, and communication." The "Pierce Arrow" provides maximum horsepower with minimum effort, always with the best possible results in mind. As with most contemporary fire fighting vehicles, models come in a variety of sizes for the greatest possible choice, ranging from the basic 750 gpm to the monster 1,500 gpm.

1: **Red is not always the colour of fire engines, as shown by this orange 1993 Pierce Arrow.**

2: **This 1996 Pierce Saber produces 1,250 gpm and comes in a variety of colours.**

3: **The Pierce Quantum of 1996 (1,500 gpm and 40 foot ladder)**

4: **The aptly-named Pierce Dash, 1987.**

5: **The Pierce Lance of 1995, including a 105 foot ladder.**

6: **1997 Pierce Arrow, including a 75 foot snorkel.**

7: **The 1997 Pierce Saber.**

OLD MEETS NEW

On Sunday, March 23, 1997, eight MK3 Scania Pumpers were brought together outside the original No. 1 station in Victoria, Australia, to reproduce a scene acted out 57 years ago and to celebrate the advances in fire engine technology over the past five decades. In 1940, eight Morris brand appliances were lined up outside the station to capture their grace and style on film for posterity. Almost 60 years later, the engines are bigger and more powerful and their integrity remains intact. The Mark III Scania Pumpers' tanks enclose 1,360 liters of water and 380 liters of foam. They are capable of pumping more liquid per minute than the old Morris Commercial appliances could ever have imagined. And they are as striking today as the Morris engines ever were.

1: **Considered the best in current fire engine production, Emergency One, also known as E-1, continue to produce some of the top machines, such as this 1990 Hush.**

Contemporary fire fighting goes beyond pumping water, although that is still central. Fire departments worldwide want complete machines, prepared for anything. No other company meets that need better than Emergency One, also known as E-One. This company operates out of Ocala, Florida, and represents the cutting edge of contemporary fire engine building. It was established in 1974 and within only 20 years became the market leader in one of the world's most competitive industries. E-One customers find the broadest ranges of vehicles covering every fire fighting contingency. There are at present over 16,000 E-One engines in active service around the world. E-One vehicles include rescue transports, quick attack units, commercial and custom pumpers, tankers, rescues, hazmats (trucks equipped for hazardous materials), command vehicles, industrial trucks, aerial ladders, aerial platforms, and aerial rescue fire fighting vehicles. In each category, there can be anywhere up to nine models! E-One remains at the forefront of modern fire vehicle design and, looking through their catalog of machines, it is obvious why. A wide variety of choices contained on a "one-stop shopping" environment gives each potential buyer exactly what they're looking for at their fingertips. It is this ease and accessibility that makes E-One so appealing and so successful. E-One is among the top fire engine manufacturers of all time, but the method by which it has achieved its success tells us a lot about where fire engine manufacturing is heading in the future. Fire fighting vehicles still have a certain charm and vitality, but all too often they have been reduced to a "product line" or a collection of expert parts. The romance that surrounded the fire fighting vehicles of the past is slowly dissipating, but the heroism and dedication of these machines and the people operating them lives on. Despite its "commercialization," fire apparatus continues to save lives and remains instrumental in our battle against fire. In fact, it is more essential than ever before, as the challenges facing contemporary fire fighting machines have become more challenging than ever.

2: **1989 E-1 model Cyclone.**

3: **E-1 Hush from 1991, with a powered Gurney lift.**

4: **The Titan (1988), one of many models offered by E-1 in their wide range of unique models.**

5: **1992 E-1 with 105 foot ladder.**

6: **This 1989 E-1 Hush included rear engines. The safety and comfort of fire fighters is now a priority for fire engine builders, and this is one step in that direction.**

7: The 1989 E-1 offers 1,500 gpm with relative ease.

8: The 1992 E-1 Hush.

9: A 1994 E-1 with 1,500 gpm and foam tender.

7

Today's truck has to face untold dangers and contain them as quickly as possible. Toxins, poisonous gas leaks, and other dangerous materials are as likely to spring up as a fire and the apparatus must be prepared to respond accordingly – the average combination fire engine used today, for example, encompasses more rescue equipment than was ever carried by the original rescue trucks. Such innovations as the aforementioned hazmat truck and the attack or mini pumpers are a response to the need for more specific equipment. Attempts to reduce the noise produced by heavy engines has led to interesting developments in engine positioning and experiments with sonic vibration to combat vibrations of air generated by an engine, thereby countering the noise. If these experiments work, they may result in whisper-quiet rigs rumbling down the streets on their way to a fire, as well as cutting down on the wear suffered by the largest of engines – economic realities are still a powerful motivator for engine manufacturers. Increased automation among fire fighting machines is becoming commonplace and the future suggests we could be looking at completely automated fire fighting systems within the next 20 years. We cannot be certain what the future holds, but we can be sure that it will be as foreign to us as today's machines would be to someone only 100 years ago.

8

PHOTO CREDITS

Picture Research:
Melissa Chapman, Keith Ryan

Main Cover photo: Lionel Martinez

Additional Photography:

CIGNA Museum & Art Collection: pp 1, 3, 5, 8, 12, 18-19, 28, 30-31, 34-35, 38-39, 40 (top), 41, 44 (top & bottom), 45, 46-47, 50, 52-53, 54-55, 56-57, 58, 60-61, 62-63, 64-65 (top), 67, 72-73, 74-75, 76-77, 78-79, 86-87, 88-89

Keith Waterton (Cameo Photography): Back jacket (top second from right and main photo) and pp 10 (top left), 13, 14, 15, 17, 20-21 (left), 24, 26, 29, 32-33, 48-49, 64 (top), 80, 82, 83 (left), 84-84, 91, 119 (right), 120 (left), 127, 128, 129 (top)

Bill Hattersley: Back jacket (top left and right), front jacket (second from right and right) and pp 2, 104-105 (bottom left), 108-109 (bottom), 111 (right, nos. 7, 8, 9), 118 (middle left), 120-121 (top)

Mack Trucks Historical Museum: pp 102, 103, 124-125, 174, 175, 176 (top left)

Imperial Publishing - A Division of Murray Cards (International) Ltd: pp 27 (bottom), 40 (bottom), 46 (left), 65 (bottom), 66 (top), 85 (right), 90, 98, 126 (left), 129 (bottom)

Shaun P. Ryan: pp 92-93, 110-111 (bottom & middle), 115 (middel right), 117 (top), 140-141 (top)158-159 (top right) 164 (bottom, no. 3), 166 (bottom left), 167 (bottom right), 170 (top left & middle), 182 (bottom left & middle), 183 (top, nos. 4, 5), 186 (top right), 100-101 (top), 113 (top right)

John Toomey: Front jacket (top left) and pp 107 (top & bottom), 108 (left), 110-111 (top left & bottom), 112-113 (top), 113 (middle & bottom), 114-115 (top left), 115 (bottom right), 116 (top), 116-117 (bottom), 118 (top left & bottom left), 118-119 (top & bottom), 140 (bottom left), 140-141 (bottom right)142-143 (top), 148-149, 150-151, 152 (top left & bottom left), 152-153 (top right), 153 (bottom right), 154-155 (top right), 156-157, 158-159 (bottom), 160-161 (top), 162-163 (top and bottom, left), 163 (top & bottom), 164 (top, no. 2), 164 (bottom, no. 4), 164 (bottom right), 166-167 (top), 168 (top), 168 (bottom left), 174 (top left), 176 (bottom left)

Bill Tompkins: pp 6-7, 9, 11, 138-139, 147, 172-173, 136-137

VOLVO: pp 178-179, 180-181

Dennis: p 113 (left)

Frank Degruchy: p 86 (top left)

Garry Kadzielawski: Back jacket (second from left), front jacket (second from left) and pp 100, 142-143 (bottom, left & right), 144-145, 146, 152 (left middle), 170 (bottom & top), 176-177 (top), 183 (bottom, nos. 6,7), 185 (top right), 186 (bottom), 188-189

American Museum of Fire Fighting: pp 16, 21 (right), 27 (top), 36 (bottom left)

Maryland Museum of Fire Fighting: pp 36-37 (top), 42-43, 70-71

Victoria Museum (Australia): pp 59, 184, 185

London Transport Museum: pp 122-123, 130-131, 132-133

Joel Woods: pp 96-97, 104-105 (top), 155 (top)

Charles Madderon: p 100

Aknowledgments
The author woud like to thank the following for their kind help and assistance: Melissa Chapman, Walter McCall, David Purcell, Keith Waterton, Keith Webling, London Transport Museum, London Science Museum, Keith Child, Andrew Hawkes, E-One, Dave Darby, Pierce Manufacturing, Volvo, Don Shumaker and Snopwy Doe (Mack Trucks Historical Museum), John Rodwell and Roy Still (London Fire Brigade Museum), London Fire Brigade, Ed Pores, Sheila Buff, Karen Del Principe, Lionel Martinez, Bill Tompkins, Steven Heaver Jr., Tony Meyers, Gene Lichtman, Joel Woods, Bill Hattersley, Shaun Ryan, Charles Madderon, Garry Kadzielawski, John Toomey

INDEX

INDEX